UNDERSTANDING
C A T S

UNDERSTANDING
CATS

ROGER TABOR

David & Charles

In memory of my old cat 'Mr Jeremy Fisher', who was almost 25 years old when she died; in celebration of Leroy and Tabitha, my present moggies; to record my congratulations to my parents on their 50 years of marriage; and to thank them for bringing me up in a household with cats and dogs, so that I could gain some understanding of both.

VIDEOS: In addition to his BBC TV *Cats* series and *Understanding Cats* series, the author has made a number of educational videos on cats with Cats on Film/Avalon Productions: *Understanding Your Cat; Breaking Bad Habits for Cats; The Cat Outdoors; The Cat Indoors; Complete Cat Care* and *The Mystery of the Cat.*

AUDIO PACKS: He discusses the neutering of feral cats and wildlife predation by cats in the 'Focus on Ferals' seminar audio pack (Alley Cat Allies, USA, 1994).

PICTURE CREDITS

Roger Tabor 1, 2, 7 top right and top left, 8 right, 9, 11, 12, 13, 14, 15, 16,18, 19, 20, 21, 22, 23, 24 bottom, 25, 26, 27, 28, 31, 32, 33, 36, 37, 38, 39, 40, 41, 42, 43, 44, 45, 48, 49, 50, 53, 54, 55, 56 top, 58 top, 59, 60, 61, 64, 66 bottom, 68, 69, 70, 71, 72, 73, 74, 75, 76, 77, 78, 79, 84, 85, 87, 88, 89, 90, 91, 92, 93, 94, 95, 96, 97, 98, 99, 100, 101, 102, 104, 106, 111, 112, 113, 114, 115, 116, 117 top, 118 top, 121 top, 122, 126, 127, 129, 130, 132, 133, 134, 136, 139, 141

Liz Artindale 5, 7 bottom left and bottom right, 8 left, 10, 17, 24 top, 30, 34, 46, 47, 51, 52, 56 bottom, 57, 58, 62, 63, 65, 66 top, 80, 81, 82, 83, 107, 109, 110, 117 bottom, 118 left, 119, 120, 121 bottom, 124, 125, 128, 131, 135, 137, 138, 140

Judy Zukoski 105

A DAVID & CHARLES BOOK

Copyright © Roger Tabor 1995
First published 1995

Roger Tabor has asserted his right to be identified as author of this work in accordance with the Copyright, Designs and Patents Act 1988

A catalogue record for this book is available from the British Library.

ISBN 0 7153 0308 2

Designed by Chris Leishman Design
and printed in Italy
for David & Charles
Brunel House   Newton Abbot   Devon

Contents

Foreword

The key to understanding cats is to look at the many aspects of the animal and to develop an integrated view. This approach throws light on the current situation, for while written accounts over the centuries tell of a growing relationship between humans and cats, the biological focus sees domestication as a successful survival ploy exploited by the cat.

In my book *The Wild Life of the Domestic Cat* I concentrated on feral cats and territories, while in my BBC *Cats* book and series I followed their historical development from earliest times. In this book and in my *Understanding Cats* series I take these approaches and

put them in the wider perspective of the practicalities of living with cats from the development of kittens to the problems associated with confining cats. Yet here too there is an evolving story. The animal that has been thought of as both divinity and demon has retained its inheritance of independence, but now in its role as pet is being increasingly restricted by fad and law. Consequently

our attitudes towards cat behaviour, breeding and care
are in a state of flux.

If you perceive a cat only as an endearing creature it
can be hard to appreciate
why it has brought out
extremes of love and hate
across the centuries. At a
time when the cat has
become the most popular
animal companion in Britain
and the USA, are we now in
danger of attempting to
redesign perfection? The cat
is not just the concern of cat
breeders, cat welfare
organisations, or pet food
manufacturers, but of us all.
Yet our close relationship

with the cat as pet in its
present widespread scale is
only of recent origin. I hope
that by displaying the merits
of the cat by the linked
media of TV, video and
book I can help towards a
greater appreciation and
understanding of cats, those
most enigmatic and
wonderful of animals.

Origins – from wild to domestic

When your cat looks at you it can convey the essence of contentment and trust, yet a second glance can reflect a little of that wildness which mirrors the rest of the family. For its size its range of weapons are as formidable as those of the tiger. As hunters, the cat family is unsurpassed. Despite domestic living, our pet cats have retained many features of their wild ancestry. This apparent contradiction makes our cats the most truly enigmatic of animals. The cat retains an independence which is treasured by all true cat lovers.

It is intriguing that the two animals that most closely share our lives as pets today, the cat and dog, are both carnivores and hunters. Yet while the dog was one of the first domesticated animals, the cat was one of the last.

This was not by chance but a consequence of the differing lifestyles of the animals, and our lifestyle at the time. As herd hunters our activities overlapped those of the wolf packs, and their ability to scavenge allowed them to haunt mankind's camps. The rest is history!

It took our dramatic lifestyle and landscape changes over thousands of years to domesticate the cat. The development of agriculture and subsequent settlement and urbanisation which arose from agricultural surpluses, produced the waste that provided scavengeable material in sufficient abundance to support a population of normally lone-hunting cats. When we see feral or gone wild cats scavenging to survive, we are watching a replay of the events that drew the cat into our lives. The move by the cat into domestic living has been a survival ploy that has worked spectacularly. While originally the cat had only a local distribution in the Middle East, it is now worldwide in huge numbers.

Both cat and dog have had a working relationship with us for most of their history. The status that most household cats hold as pets today is a very recent role. We are still adjusting to that changed role. Its significance should not be underestimated, for numbers and distribution of all other cat species are in decline due to habitat destruction and conflict with man.

It has been during the period that the domestic cat moved into predominantly pet status, that the decline of wild cats has been so dramatic. From the mid-

Feral cats scavenging from a rubbish bin in Spain. In urban areas scavenging is more productive for feral cat survival than hunting

nineteenth century mankind's own numbers and destruction of huge areas of virgin planet surface have exploded exponentially. As man thrived, so did the domestic cat due to the massive increase in food supply for both house and feral animals.

How Did It All Begin?

Ancestors of the mammalian families of today began to emerge during the reign of the dinosaurs. Their temperature control allowed them to move at night and to survive cooler conditions than reptiles. Developing from these 60 million years ago, the Miacids, a group of forest-dwelling, flesh-eaters flourished, which gave rise to all the groups of today's carnivorous mammals. Around 50 million years ago Dinictis, an early form of cat about the size of a Lynx, emerged and today's cats developed from it. The first big group of Felidae to become established were the Sabre-toothed cats around 34 million years ago, when many large mammals developed as the climate cooled. At first sight, the Sabre-toothed cats seem a crazy spur for evolution to have taken, for surely those enormous teeth would have been a great hindrance to eating by blocking the

Compare the skull of a modern Tiger (foreground), with that of a Sabre-toothed cat, Smilodon. The length of the Tiger's stabbing teeth of the lower and upper jaws combined is similar to that of the Sabre-tooth

Opposite: The Cheetah has proportionally the smallest canines among cats

mouth! However, they are one of the cat's great success stories, for the Sabre-tooth was the dominant type of cat from the Miocene to the end of the Pliocene. Some Sabre-toothed cats were still about only 13,000 years ago, so they survived as a sub-family for nearly 34 million years, while in contrast many modern cats and mankind have been around for a brief time.

The spread of grass landscapes around 25 million years ago encouraged huge grazing herds, benefitting the lion-sized Smilodon, the biggest Sabre-toothed cat. Smilodon could not use the asphyxiating neck bite, or the modern cat's neck bite with open mouth, but it could stab prey to kill it.

The line which gave rise to our present cats developed around 15 million years ago. The demise of the co-operatively hunting Sabre-toothed cats happened with the extinction of their large ungulate prey 10,000 years ago. The dominant role of the Sabre-toothed cats for such a long period in the development of the cat family

underlies the absurdity of calling the big fangs of the cat 'canines'! On any species of modern cat, even without the Sabre-tooths, these teeth are far larger than in the dog family. They really should be 'felines'!

The Modern Cat Develops

Today's cats both big and small, although having large canine teeth in the upper jaw, also have a pair of near matching 'daggers' in the lower jaw.

Biochemical sampling shows the South American small cats branched around 12 million years ago in the cat family tree. About 9 million years ago the ancestors of the small cats of the Old World developed. In this analysis the great cats and the mid-sized cats (including the Puma and the Cheetah lines) became distinguishable around 5 million years ago, with the Lynxes diverging from the great cats only 3 million years ago.

The placing of the Cheetah in the pantherine lineage from DNA and other biochemical data came as a surprise. The strong specialisations of its lifestyle and appearance had led to a belief that it must have evolved separately early on.

The Cheetah is the cat farthest from the Sabre-tooths in having small canines to allow for the larger nasal opening that enables it to increase its air intake during a high speed chase. However, biochemical evidence needs to be considered alongside that from paleontology and other areas. At present the cheetah's taxonomic position is still a matter for conjecture.

The Cat Family – Taxonomic Tangles!

It is far from clear how many species of modern cat exist, as taxonomists continue to change what they consider is a species or subspecies with regard to certain groups of cats. The Scottish Wildcat was once considered distinct from other European Forest Wildcats, then all of the closely related forms across not only Europe but down to the African Wildcat came to be considered races of one species. Not everyone is in agreement.

This is not limited to individual species, but also to the main divisions. Most taxonomists put the majority of the big cats into *Panthera*, except the Cheetah, which is put on its own in *Acinonyx*, the Clouded Leopard in *Neofelis*, and the small cats into *Felis*. But some put the Clouded Leopard into *Panthera*, while others take the Marbled Cat from the small cats and put it into *Pardofelis*, and put that into the *Panthera* too – and so on!

Non-biologists usually give up trying to make sense of cat classification when they realise that the Puma, which any tape measure would put into the big cats, is put into the small cats because of a tiny difference in the hyoid bones and larynx in its throat!

R.F. Ewer, just twenty years ago, could write 'the nomenclature of the Felidae presents a major problem and almost the only point on which there is universal agreement is that the Cheetah requires a genus of its own.' Even that is disputed now.

However, the Ocelot, Margay, Tiger

A Siberian Tiger, the largest of the modern cats

Cat and Geoffroy's Cat are all bracketed into the subgenus *Leopardus* as they have thirty-six chromosomes rather than the thirty eight of most cats, including the Domestic Cat.

CATS IN CRISIS

The days of big animals in the wild are numbered, the days of big carnivores more so, and those of the solitary hunting forest cats particularly. The reason is simple – they require huge areas for home ranges in which to hunt. Even without other pressures such continuous habitat is becoming increasingly fragmented and destroyed.

The immediate story of the big cats over the next thirty years will be the contrast between those requiring specific habitat, like the Tiger and Jaguar, and the generalists, such as the Leopard and the Puma. The specialists that require forest habitat are doomed in comparison to the generalists who will probably survive though at lower levels than today. The specialists will only remain in zoos or be sustained with great effort in special specimen reserves.

The Great Cats' Survival – The Tiger Test Case

Officially the Tiger has been saved, but unofficially its collapse is inevitable. It has been brought back from the brink of destruction mainly by the actions of 'Project Tiger', involving the World Wide Fund for Nature, the International Union for Conservation of Nature and various governments, particularly that of India.

During the nineteenth and twentieth centuries shooting parties made up of Maharajas and the British Raj, followed by a ruthless, post-Independence agricultural policy, drastically reduced the entire Tiger population of India to under 2,000 by the end of the 1960s. Worldwide, only 5,000 Tigers remained when just forty years before there had been over 100,000. Firearms, agricultural poisons and habitat loss by massive deforestation across Asia eradicated 95 per cent of Tigers in half a lifetime.

The situation is worse than is usually realised, for entire races of Tiger have already gone. There were eight races, but in 1937 the Balinese became extinct, followed more recently by the Javanese. Horrifically, both the Chinese and Russian governments attempted to exterminate entire Tiger populations in the name of agricultural productivity. Forest clearance and direct Tiger killing by the Russian army probably sent the Caspian Tiger to extinction. China's extermination policy almost eradicated the Chinese Tiger, but fortunately in 1979 official policy changed to protect it. The effects of war in Laos, Cambodia and Vietnam, and military rule in Burma, have been disastrous for Tiger populations. The Siberian Tiger has been badly mauled for the Far East medical trade.

The response in countries like India to dwindling numbers has been magnificent. Tiger reserves have been set·aside and numbers have increased. Yet the key problem remains that in India most Tigers are in small isolated populations doomed not to survive.

What is needed is an adequate gene-flow, which will require close monitoring and co-ordination. Due to the active-breeding sex ratios, it is the male lines that are restrictive. Moving adult and sub-adult Tigers around will

be fraught with behavioural and territorial problems. Capture and release of the animals, plus artificial reproduction techniques, will be more likely to succeed. The use of semen, eggs and embryos from zoo animals will be of the greatest benefit – provided that the Tiger races and original areas of the animals are known.

Species survival computer programmes are currently employed in zoos for big cat pairings to ensure that races survive in captivity. While simple artificial insemination in wild cats has physiological problems, both embryo transfer and in vitro fertilisation offer

This Bengal Tiger in north India is confident that its camouflage markings blend it into its surroundings of tall grasses; but such wild habitat is disappearing fast

more promise to isolated wild populations. These techniques were first used successfully on the Domestic Cat in 1988, and since then they have been used on Tigers and other wild cats in zoos. The next step is to support Tigers in the wild. These artificial reproduction techniques could prevent the genetic collapse of not only the Tiger but also other endangered cats.

Other Big Cats

Loss of distribution range plus loss of numbers has been a consistent feature for cats in modern times.

Most people believe that the Lion, as 'King of the Beasts', is secure, but few monarchies are secure today. Seven races or subspecies of Lion have been recognised, but the last southern Cape Lion was shot in 1865, and the northern Barbary Lion became extinct in 1922. Until recent historic times Lions ranged across Africa and throughout the Middle East and North India. However, firearms shredded these populations so that by World War I only twenty Lions remained in India's Gir Forest. Although these have increased, their low ebb led to fertility problems.

Like the Lion, the Cheetah's range encompassed India as well as Africa, but the last three Indian males were shot in 1948. The Cheetah cannot cope with habitat changes, and between 1955 and 1970 the world population halved. Due to population losses in the past from which the Cheetah recovered, their genetic base is narrow, making them susceptible to disease.

The Jaguar once ranged across the south west of the USA and throughout Central and South America. Now its only stronghold is in the Amazon Basin, and forest destruction has put it under pressure.

The huge international trade in wild cat skins, in particular the spotted cats, has been massively reduced by the CITES agreement, but the trade still flourishes.

The Small Cats

Not having the 'pin-up' status of the big cats, the future of the small cats is less secure. They are featured less in zoos, so breeding programmes are less established. Being elusive in the wild, their natural status is often unknown. We hardly know a thing about some small cats, including numbers.

For example, very little is known about the Kodkod, the Andean Mountain Cat or the Borneo Bay Cat. We know little of the Oncilla, apart from

The Pampas Cat: little is known of this South American cat's breeding patterns

the increasing numbers killed for their skins because other spotted cats have declined. There are big gaps in our knowledge of the Flat-headed Cat, the Pampas Cat, the Rusty Spotted Cat and even the Clouded Leopard. The Chinese Desert Cat does not live in deserts, it is still unclear what type of cat is the Iriomote, and what is the Onza? It is likely that a number of these cats will pass into oblivion without our knowledge of their lives improving very much unless supreme efforts are made.

Fortunately, not all small wild cats are in the same situation, but that does not make them secure. The main ancestor of the Domestic Cat, the Forest Wild Cat (*Felis silvestris*) has a range that encompasses Britain, Europe and a large part of Asia, and in the form of the African Wild Cat (*Felis silvestris lybica*) takes in much of Africa. Agricultural changes, woodland loss and persecution in Britain and Europe robbed it of much of its distribution.

THE EGYPTIAN ORIGINS OF THE DOMESTIC CAT

The nearest living relatives of the first domestic cats can be seen in the centuries-old Khan el Khalili bazaar in Cairo.

These lithe, self-sufficient cats that silently patrol the ancient alleyways are direct descendants of the first of their kind. They are fed by shopkeepers, who continue a long tradition of benevolent attention to street cats in the city.

The Domestic Cat was a species born into this world with the status of a god. This was not just recognition of the

'other-worldly' character of the small cat, but because of its identification with its larger cousins.

From earliest times in Ancient Egypt, a link existed between the sun's searing heat and the aggressive power of the lion. The lion-bodied sphinx guards the pyramids at Giza and faces due east to meet the sun's first rays. The Pharaohs Cheops and Chephren, who built these pyramids, also built the earliest part of the temple of the cat goddess Bastet at the ancient city of Bubastis in the Nile Delta around 4,500 years ago. It was then dedicated to a lion-headed Bastet, similar in appearance to Sakhmet. Bastet's link was with productive warmth, while Sakhmet's was with desiccating heat. When the small cat became domesticated, bronzes and carvings of Bastet began to be made as a small cat or as a cat-headed woman. Bastet was a local deity until a Libyan chieftain, who had made Bubastis his base, seized the throne of Egypt. He made his city the capital, and its

The Fishing Cat is endangered in parts of its range. It lives in wetlands where it scoops fish out of the water, and even dives after them

A bronze mask of an Ancient Egyptian cat mummy

goddess took national prominence. In this city of cats, the goddess' temple was built of finely worked granite, but today not one stone remains standing upon another.

Herodotus, who visited Bastet's temple in the fifth century BC, when the cult of the cat was at its height, recorded that it was the chief annual assembly for a god. Over 700,000 people made the pilgrimage, which must have been the bulk of the country's population.

It is worthwhile trying to visualise the scale of events and its importance to the participants, for archaeologists have invariably marginalised the importance of the cat preferring the lure of excavations of the pharaohs. Yet to contemporaries, the cat goddess was the key female fertility deity, on an equal footing with the Greek Artemis or the Roman Diana. Millions of cats were mummified and buried at the extensive cat cemeteries of Bubastis and at other cat cult centres.

We know from Herodotus' own words that the festival dedicated to the cat was celebrated 'with abundant sacrifices' and X-rays have shown that a number of cats had their necks broken. Yet when a cat died naturally, householders shaved off their own eyebrows in remorse, and carried their cat to Bubastis for embalming and burial. This apparent contradiction of sentiment is not unusual, for most religions held an element of ritual sacrifice of an aspect of the deity.

Despite the huge deposit of material which remained buried intact for over

3,500 years since domestication, most mummified cats were discovered and mined out in the nineteenth century. After a handful of the grander cat mummies reached the main museums of Europe, their economic value fell. Most ended up as fertiliser and shipments even reached Britain for this purpose. After this storehouse of information had been kept for centuries, before we could work out the true origins of the Domestic Cat, the material was

Top: A seated cat, as Bastet, chiselled in sheet gold on Tutenkhamen's funery case

Above: The mummified head of a mature adult cat, still retaining its skin and some hair, from Pakhet's Temple. The skull of a modern cat lies behind it

Opposite: A street cat in the Khan el Khalili bazaar, Cairo, identical to the first domestic cats. There is a centuries-old tradition of feeding these stray cats in the city

A feral cat on one of the limestone blocks that make up the Great Pyramid at Giza, Egypt

Opposite above: The African Wildcat, the main ancestor of the Domestic Cat

Opposite below: The Marsh Cat is more tractable than the African Wildcat

destroyed. Due to the importance of finding the origin of the Domestic Cat, I pursued it in Egypt.

While Bastet's city has been encroached upon by the spreading urban sprawl of modern Zagazig, the rock-carved Temple of Pakhet, the cat goddess of Middle Egypt, remains isolated. The earliest domesticated cats had once walked in its carved-out rooms, and it may be where a significant part of the story of the cat's domestication took place.

While it has been held that the only ancestor of the Domestic Cat was the African Wildcat (*Felis lybica*), I believe I was the first to identify the Marsh (or Jungle) Cat (*Felis chaus*) in one of the earliest cat paintings in Egypt in Khnumhotep's tomb, close to the temple. It is most unlikely that the artist could have made such a sensitive portrait of this elusive animal without the familiarity gained from its captivity. In the temple's Ptolemaic extension, I found bones of the Marsh Cat, showing they had been kept captive alongside the African Wildcat.

It is possible that domestication took place just by more tractable individuals of the African Wildcat exploiting the scavengeable wastes of the developing towns. Yet the confining of the two species of wildcat in close captivity would have produced hybrids and that may have been critical in the initial stages of the domestication story.

Amazingly, during the filming of my BBC TV series *Cats* at Pakhet's Temple, I discovered only the second ancient Egyptian cat cemetery to be found this century. It was thought that the area had been completely mined out in the nineteenth century and on my previous searches I had only found a scattering of bones. But as we started filming, I suddenly realised that overnight a 9-nch seam of cat mummies had become visible as a result of movement on a slope of loose material fronting the temple area.

All along the 200-foot seam skulls and bones protruded, there were limbs still covered with fur, and bandages brown with age. From this site, where both Marsh and African Wildcat were kept, it may at last be possible to work out the origin of the Domestic Cat.

The best-bred cat – the moggie

A Mackerel Tabby in Turkey, of the slim form found around the Mediterranean: the original moggie

Opposite: For centuries the cat was a working animal, controlling rodents, rather than a pet. This English Tabby, seen in a water-mill, still has the original spotted mackerel pattern, but it is of 'cobby' northern build

Moggies are good, gutsy animals that have the best possible pedigree going back unhindered through the mists of time. Because of their near random mating they have enjoyed the best possible breeding programme for countless generations. Their survival and subsequent breeding derived from natural selection, favouring functioning characteristics. It is an unbeatable combination for healthy animals, producing a natural perfection of design and function.

Unfortunately the media carries out an unwitting 'conspiracy' against the poor old moggie because of the need to *see* variety. Far more people live with moggies than breed cats – 90 per cent are moggies – yet picture editors are not alone in liking to see variety on their pages.

The high-profile events that promote cats – the major cat shows – primarily promote breeds and breeding, rather than moggies and selection of good pets. Of course, breed cats can and do make good pets, but as millions of people will testify, so do moggies.

I admit to a bias because my cats are moggies. I choose to have moggies as I believe they are proper cats, the essence of cat and not genetically messed about. Among cat writers this must put me into a fairly serious minority, for traditionally most writers have been breeders with a bias not only towards breeds, but their special breed. My approach as a cat biologist and behaviourist is fundamentally to look at this incredibly successful species that shares our lives, and find out what makes it tick – to try to understand the cat.

Moggies on average live much longer than breed cats as they are genetically more robust. Over recent years breed cats seem to be reaching better ages than before, but usually with significant veterinary attention.

In the real world the majority of cats are moggies, with pure breed cats a tiny minority. Areas of geographical isolation have historic breeds as street and house cats unconnected to the cat fancy. In some countries the local moggie population is made up of the original cats of a cat fancy breed, such as the Angora in Anatolia, the Siamese and Korat in Thailand and the Japanese Bobtail in Japan (it took the occupying Americans after World War II to point out that the Bobtail was special). In the same way it is easy to underrate the real British, European and American Shorthair on the streets, for to us it is just 'the moggie'.

STRIPES AND SPOTS

In its journey north from Egypt, the Domestic Cat entered the realm of the European Forest Cat. Just as today the Scottish Wildcat is hybridising with feral

Domestic Cats, so when any of the new domestic animals became lost and turned feral, they would have mated with wild cats. It is hard to believe that the fuller figure of the British and European moggie came into existence any other way, although it could have happened by mutation. Compared to its African ancestors, the bolder markings of the moggie are consistent with such hybridising. The original slim, paler cats are still to be found around the Mediterranean basin where the climate suits their build, and they remain as the southern street moggie. Among the

THE STREET CAT
The moggie is the house cat, the pet, the street cat, the farm cat – the everycat. In Britain and Europe, North America, Australia and New Zealand the cats have a common European solid build. In the Mediterranean, southern Asia and South East Asia, the warm climate has kept the build closer to its Egyptian origins.

The Blotched Tabby cat is darker than the striped form, so was able to merge safely into the shadows of European towns during the witchcraft era

early Egyptian cats were both fine striped and spotted tabbies, and they are still there.

As there is a gradation in tabby markings of striped into spotted, the cat geneticist Roy Robinson has suggested that instead of the spotted and mackerel tabbies being from different genes, it is more likely that they are one type modified by polygenes (see p48). While this is probably the case I think this part of their inheritance occurred before domestication, for the African Wildcat has both a spotted and striped form and Egypt is located centrally to their ranges. The genetic reinforcement by the Domestic Cat hybridising with the wild spotted eastern variety when it was taken to India, may well have caused the preponderance of beautiful blue silver spotted tabbies in Nepal.

Tabby Types

All cats are genetically tabbies, even if they do not look like it! Even a black

and white cat carries tabby genes. There are three types of tabby patterns: the striped 'Tiger' or 'Mackerel' Tabby, the Blotched Tabby and the Abyssinian form.

In early medieval Europe a new type of tabby appeared – the Blotched Tabby. Instead of the pattern of narrow parallel bands seen in the Mackerel Tabby, the new mutation had blobs and swirls. The new tabby spread round the world, carried on ships travelling to various parts of the growing British Empire, and so a number of years ago I dubbed the humble Blotched Tabby 'the British Imperial Cat'. The Blotched Tabby has not only reached as far as Hong Kong and Australia, but is also surviving in the freezing conditions of sub-antarctic islands alongside penguins.

Only in recent years has it been possible to trace the spread of marking and coat colour mutations as researchers in different countries pooled their observations to draw up gene frequency maps. These show not so much the distribution of the type of cat (the phenotype) as of the mutated gene itself (the allele). In 1977 Neil Todd found by plotting the frequencies for the allele for the Blotched Tabby that the initial waves of settlers sailing from Britain to America in the seventeenth century, Canada in the eighteenth century and Australia in the nineteenth century took increasing numbers of Blotched Tabbies among their cats as each century passed. This happened as the proportion of Blotched Tabbies increased in Britain. Their increase probably occurred because the Blotched Tabby has a darker coat than the Mackerel Tabby. From medieval times until the nineteenth century, landlubbers, unlike sailors, had little time for cats. Consequently, the darker coat of the Blotched Tabby was a decided aid to survival in the murky shadows of a town street. Most cats then lived alongside us rather than with us.

If the wildtype stripes of the original Mackerel Tabby are 'country tweeds', then the greater amount of black in the Blotched Tabby makes a 'city suit' cat. While 'country clothing' blends superbly in rural conditions, the 'city suit' fits urban low light conditions.

Sailors had a high regard for cats because their own survival could depend on them. Rats and mice on board ship could ruin a cargo, destroy the food supply and even endanger the ship. So while the oceans have stopped the spread of many animals, they have assisted the distribution of the cat, with a little help from man.

From the distribution of the Blotched Tabby mutation, Britain still holds the highest density of these cats, and working back from present day figures it probably occurred around AD1200.

One major type of tabby, the Agouti or Abyssinian-type, is rarely thought of

HYBRIDISING WITH WILDCATS

During medieval times the Forest Wildcat (Felis silvestris) was hunted for sport and for its fur, as was the feral Domestic Cat. It was lost from southern England probably by the sixteenth century. By World War I it had retreated virtually to the Scottish Highlands. Its eradication from England was largely due to deforestation, a habitat change that better suited the feral living Domestic Cat. In Scotland gamekeeping on shooting estates was more responsible. When this ceased after World War I the Scottish Wildcat seemed to recover very quickly and expanded into the Lowlands.

Yet when skulls were examined the average Scottish Wildcat was said to be only 66 per cent Wildcat due to hybridising with feral Domestic Cats. This has also been found across Europe with an average of 63 per cent. As the European Wildcat across Europe is noticeably smaller than in Neolithic times, this could suggest a long period of hybridising from when the Domestic Cat began its expansion.

A feral street Agouti-patterned Tabby in Bangkok, Thailand, showing the leg and face patterns that breeders of Abyssinians try to avoid

by most people as a tabby at all, mainly because it lacks stripes. However, even in the breed form, there are tabby markings on the forehead, face and tail.

Breeders have worked hard to reduce these markings and a street Agouti Tabby has far clearer face and forehead patterning.

The show Abyssinian's beautiful cinnamon agouti coat, with its 'natural shadowing' has the same natural shadowing of the Jungle Cat. It makes Abyssinian kittens perhaps the prettiest of all. Devotees suggest that it has been around since the time of the ancient Egyptians. It is arguable that some of the Egyptian statues show agouti-marked cats, but these could well be Jungle Cats which were kept captive on temple sites.

Abyssinian kittens with a bronze statue of Bastet. The clear Tabby markings on the kittens' foreheads fade as they become adult

Opposite: A black cat beside a Venetian canal: this type owes its distribution to an earlier Mediterranean trading empire, the Phoenician

A ginger Tabby parked on double yellow lines!

I have found most agouti-marked Domestic Cats to be in South East Asia, usually with grey-looking agouti coats.

Other Tabby Colours

The distribution of the gene of the ginger cat indicates that their original mutation occurred in Turkey centuries ago and was carried north by Viking traders who liked the cat's unusual colour. Travelling by rivers they took the cat to the Baltic from Byzantium via the Black Sea. The ginger toms found on Scottish islands and in Ireland are living relatives of the cats that lived in Viking settlements.

While the basic tabby has black as its dark marking, the flecked section varies from yellow, to tan and full brown. This variability is believed to be due to polygenes. A fine ginger tom is still a tabby even though the black markings have been replaced by orange ones. True tabby markings can be found under such guises as cinnamon, blue and silver colourings — and not just among fancy cats!

The Black Cat – The First of the Solid Colours

Gene frequencies suggest that the initial spread of the black cat, one of the earliest mutations, was aided by Phoenician shipping. These active traders around the Mediterranean were operating at a time when very few cats had been allowed out of Egypt. The adherents of the fertility cult of the cat goddess Bastet actively sought to confine them within their country. As it left Egypt the cat began to change from god to mog!

From altar to showbench

When the black cat arrived in Europe, it was accepted as a working cat. But as the centuries passed it was destined to become synonymous with 'the Prince of Darkness'

THE FEMININE FELINE

Soft and sensuous are descriptive words that can be applied to a cat as equally as to a woman. A fable credited to Aesop tells how a man fell in love with his cat which Venus, in compassion, changed into a beautiful woman; however, when a mouse ran by their

bed she leapt out onto it, changing back into a cat! So readily interchangeable are feline and feminine in our minds that in advertisements an elegant cat will be shown alongside an equally elegant woman to convey a sexy image.

There has been a long-held association of dog and man, and

woman and cat. Fundamentally, it may just reflect the historic role of the sexes and the associated role of these animals; man as hunters went out accompanied by their dogs of the chase, while women worked around the homestead where the cat slept beside the fire and stalked mice in the yard.

Before the rise of Christianity, all the religions across Europe had a mother earth goddess figure who was held as a fount of fertility while being eternally virginal. For the Greeks it was Artemis, for the Romans Diana and the Scandinavians Freyja. The Egyptian goddess Bastet was associated with fertility in women, and warmth, love, dance and the moon.When the Greek Ptolomy became pharaoh, he had inscriptions carved on the cat rock temple near Beni Hasan, which he called 'Speos Artemedos' – Temple of Artemis.

Diana was a moon goddess, living in celibacy and presiding over women, as did Bastet. So, not surprisingly, when Typhon waged war on the gods, Diana escaped – by turning into a cat! In London, St Paul's was built on the site of a cult shrine to Diana. A sistrum has also been excavated in London, which was used in the worship of the Egyptian fertility goddesses, including Bastet. It is an intriguing thought that worship of the cat goddess Bastet may have taken place with that of Diana on the site of St Paul's in Roman Londinium.

Despite the best endeavours of the Egyptians not only the reputation of cats leaked out of Egypt, but so did some of the animals. Illustrations of Domestic

Cats first began turning up in Italy around the fourth and fifth centuries BC. When Christianity was made the Roman state religion, in the fourth century AD, other religions such as the following of Bastet were banned.

In Egypt the main cat divinity was Bastet, whose chief centre of worship was in the city of Bubastis. At the massive annual pilgrimage each year, thousands of devotees, mainly women, sang, danced, drank and shook sistra.

The ancient Egyptians noted that the size of the pupil of the cat's eyes varied with the phase of the moon. Although this is just due to the amount of prevailing light, it was thought to be magical. As a woman's cycle has a monthly periodicity, links were recognised between women, cats and the moon.

The cat goddess' fertility role for women is revealed in tomb wall paintings found near Luxor. In scenes of domesticity, husband and wife are

Above: Bastet in the form of a cat-headed woman with four kittens at her feet and shaking a sistrum, symbolic of her role as fertility goddess

Below: Egyptian ladies and girls with a litter of kittens born in the site excavations at Bubastis, the cult centre of women's fertility dedicated to the cat goddess Bastet

portrayed as they hoped to be in their after-lives. When a cat is shown with them, it is always under the seat of the woman. The cat as an incarnation of the goddess was positioned to enhance the wife's after-life fertility.

Although the Egyptians' regard for their cats diminished through the passage of time in Europe, the historic link between cats and women persisted, and indeed has lasted until the present day.

THE CAT MOVES NORTH TO BRITAIN

The Phoenicians could have introduced the cat to Britain, but while Himilco of Carthage may have landed on an exploratory voyage around 425 BC, evidence of any direct trade is lacking.

Unambiguous evidence of a Domestic Cat appears in Camulodunum (Colchester), the main 'city' of Celtic Britain just ahead of the Roman invasion. A cat bone found alongside domestic stock is strong evidence of the domestic animal. I like to think it was a diplomatic gift to the old 'king of Britain' who the Roman authorities had been 'keeping sweet'. When he died they invaded.

Abundant Domestic Cat remains have been excavated from several Roman fort and villa sites often found in wells that had been used for rituals.

The Cat in Ireland

Once the cat had reached England, it is most likely that it arrived in Ireland as a consequence of trade. Excavations show that the Domestic Cat was not only widespread in monastic sites across Ireland during the early Christian era, but that it was enjoying a contented life, as revealed by a contemporary monk's poem about his feline companion, Pangur Ban. These cats reached a good age.

In contrast excavations on Viking and medieval towns have shown that although cats reached adulthood, most died young. The same young age pattern has been found from some English medieval cities. Cat skulls from medieval York and from an Irish site show knife marks around the orbits, which is evidence of skinning. Pelts of all sorts, including the cat, were an important part of medieval trade.

The cat's role changed from contemplative companion to a utilitarian commodity. As its numbers increased with the increase in the size of towns its rarity value declined. In 1358, Edward III bestowed a murage grant of a halfpenny on the port of Youghal in Ireland for the exportation of each one hundred cat pelts.

The Witches' Cat

When one religion usurps another, the deities of the old faith are relegated to the position of demons. When Christianity replaced the older religions, followers of the fertility deities were deemed heretical. Time amalgamated beliefs, and followers of the old fertility rituals were hounded as witches.

The woodland hunting biology of this new animal to Europe was not understood by medieval man any more than it was by the ancient Egyptians, but while shining, reflective eyes helped

VIKING FURS

Evidence that the Vikings used the pelts of the Domestic Cat in their everyday lives is revealed by the skinner's pit excavated at Odense, Denmark. Dating from AD1070 it contained the remains of at least sixty-eight cats. The cats had been killed by a sudden wrench of the head, and had skinning cutmarks. Most of the cats were about ten or eleven months old so their coats would have been adult size but undamaged. This led Tove Hatting to suggest that they had been bred in captivity to provide fur. The remains of the fully grown adults were of slighter animals than the Danish moggies of today, which is consistent with the arrival of the lighter built eastern Mediterranean cats by Viking trade.

Opposite: Initially the cat was accepted by the Church in Europe, but this attitude changed as the cat became identified with the devil

make the cat a goddess to the Egyptians, they reinforced its damnable identity in Europe.

From witchcraft trial documents, it is apparent that in Britain the cat as familiar was central to witchcraft. To accusers a cat could be an agent of the devil or the devil himself. They believed that the wishes of a witch were enacted by the demonic cat familiar against her neighbours.

The first major witchcraft trial in England took place in Chelmsford, Essex, in 1565 and set the pattern for later trials. It revolved around the ownership of a cat that was dangerously called 'Satan' by three women. An original report of the trial survives in Lambeth Palace Library. The accusers thought that the women suckled the cat with blood from unnatural teats. It became routine to search an accused woman for a 'teat', and a wart was considered damning evidence. From the trial records, neither the women nor their accusers believed that the women had any supernatural power until they obtained the cat.

Only a few enlightened independent-minded people led the vanguard of the cat's acceptance, but the country that showed the way was France. Although cats were still being burned alive in street bonfire festivals in the seventeenth century, cats became fashionable among French aristocrats at the court of Louis XIV. The leading lady poet at the court, Madame Deshoulieres, wrote poetic love letters on behalf of her cat Grisette to the cats of her friends and they replied in like manner. This was remarkable, for

The night-reflective eyes of the nocturnal hunting cat were misinterpreted as magical by Ancient Egyptians and medieval-to-eighteenth-century Europeans

in the era of witchcraft persecution, from the mid-sixteenth century to the end of the seventeenth, ownership of a cat could result in a death sentence.

Even in the early eighteenth century, in Caithness the link between cats and witches was still strong enough for the deaths of two elderly women and the injury to another's leg, to be attributed to the actions of a man who had attacked a group of catawauling cats with an axe the night before.

As 'witches' found, identification of women with cats is not always kind. Women are accused of being 'catty', but not men. Yet this is not new, for in the eighteenth century a common name for a prostitute was 'cat'.

CATS AND WEATHER PREDICTION

From all ages and all around the world,

Opposite: For centuries across Britain and Europe cats were burnt alive in wicker baskets, suspended above flames to prolong their agony, as it was believed that this hurt the devil (a reconstruction for the BBC TV series *Cats*)

Girls dressed up as cats at the annual Ypres Cat Festival in Belgium, demonstrating the interchangeability of feminine and feline

the ability of the cat to anticipate changes in the weather has been recorded. Mariners have known that when a cat becomes skittish 'it has the wind up its tail'.

A major reason for the official vindictive attitude to witches in seventeenth-century Britain stems from the severe storm that almost wrecked the ship on which James VI of Scotland was returning from Scandinavia with his new wife Queen Anne. Alleged witches claimed to have instigated the storm. James became outraged when one of the witches was acquitted and he took over the running of the trial. The women 'confessed' that they had conjured the storm by roping bits cut from a dead man to a cat, which was then thrown into the sea. He then spent years writing his book *Demonology*, which was published in 1597. When he was made James I of

England he published a new edition and instigated a draconian new witchcraft law.

James did not doubt the power of witches over weather. Unfortunately for the cat it could be demonstrated that they could foretell the weather. It is particularly attributed with anticipating rain, though fortunately today we realise that is not the same as causing it.

Robert Herrick, the English poet, was just six years old when James' book was first published. Herrick lived in isolation in a Devon vicarage where he came to delight in the continuity of pre-Christian rural customs. In his peaceful seclusion, cats were Herrick's main companions and they confirmed for him the traditional belief that cat's could predict the weather:

True calendars as Pusse's eare,
Wash't o're to tell what change is neare.

With such interests it is small wonder that when the Puritans came to power they threw Herrick out of his living.

In 1643 Herrick's contemporary, John Swan, wrote even more explicitly of the action in that the cat '…wash her face with her tongue; and it is observed by some that if she put her feet beyond the crown of her head in this kind of washing, it is a signe of rain'.

Fortunately for the cat, at a time of anti-witch and anti-feline hysteria, some kind and careful observers realised that the creatures were passive detectors.

For thirty years I have monitored the ability of numerous cats to predict rain in this way, and have yet to find an exception. It really does work! Twenty years ago, I enlisted the help of viewers of BBC 1's *Animal Magic* to see if others found the same and how long after the cat washed over its ear did it rain. It was found that most ear washing occurred within four hours of rain, particularly the last hour.

I have no doubt that the cat has information about changing weather; but if it interprets or acts upon that information is quite another matter. I see no reason why not. Avoiding rain can give a survival advantage at times.

I believe that the action of the cat in washing over its ear, as the cat on the front cover is doing, is due to the humidity and pressure changes making themselves felt upon the membranes of

Cat carved on the sixteenth-century window-sill of an East Anglian weaver's cottage, in the heart of witchcraft country

The commemorative roadside stone to Dick Whittington and his cat on Highgate Hill in London where, according to tradition, he 'turned again' on hearing Bow bells

the inner ear and eardrum. It must feel something like the air pressure changes we undergo on take-off in a plane.

THE WORLDLY STAGE OF PANTOMIME CATS

In Britain at Christmas-time generations of children are used to the sight of a young woman, dressed as a young man, with a huge cat cavorting on the stage beside her. It's that curiously British annual event of pantomime, when the legend of Dick Whittington and his cat has Bow Bells say: 'Turn again Whittington, Lord Mayor of London!'

The pantomime, one of the stage's most enduring forms of entertainment, grants an intriguing view of the cat across many centuries. During the height of its popularity in the nineteenth century, the cat was still gaining acceptance after the evil slur of witchcraft, yet the image portrayed is often based on an earlier one of powerful magic. Pantomime helped to make cats more acceptable, assisting the change in attitude which led to the setting up of the cat fancy in 1871.

The Lord Mayor and his Cat

In the story of Dick Whittington, Dick and his cat return to London and go to sea with his merchant employer to a sultan's palace overrun by rats, which the cat kills. Dick is rewarded with a fortune and becomes Lord Mayor. Sir Richard Whittington really was Lord Mayor in 1397, 1406 and 1419, and he did make a fortune – but how?

The pantomime derives from a play in the mould of the tales of French magic cats, where the cat aids its master with good fortune. *Puss In Boots*, also written down in the seventeenth century from traditional tales, also has a poor youth made rich by his cunning cat which tricks an ogre into turning into a mouse. Having a cat for 'good luck' was not passive but in some way thought to cause the luck, but there is no evidence that Richard Whittington ever owned one.

What of the rats in the legend? Dick was born ten years after the Black Death had wiped out half the population of London. At the time the link between rats and disease was not known, but the black rat population was enormous and the cat's controlling role was understood. Belief in the cat's magical powers led to the creature's interment in house walls to keep rats and evil away.

Similar stories to Dick's exist – back to a tenth-century tale from Persia – suggesting that the legend was grafted to his name once he had been linked to sudden good fortune and a cat. Amazingly, as in the pantomime, Dick really married Alice, the merchant's daughter. The medieval sense of humour may have been at work on the Lord Mayor, for a clue lies in the two last lines that were inscribed on his tomb:

For lo! he scorned to gain by stealth
What he got by a cat'.

Dick gained a specific fortune by a cat rather than steal it. The 'cat' was Alice, and when he married her, he suddenly became much richer! 'Cat' was a common term for a spiteful woman. The old expression 'to live under a cat's foot' described a henpecked man. Dick may have married his 'cat' for 'good fortune' and needed a bit of magic for that match! The tales of *Dick Whittington*, *Puss-in-Boots* and others are the positive side of 'cat-magic' seen by the perpetrators rather than the persecutors.

The White Cat

A very popular nineteenth century pantomime that has almost disappeared without trace was *The White Cat*. The heroine, a cat-headed woman, looked remarkably like the cat goddess Bastet. A king's three sons had to fulfil tasks to inherit the crown. The youngest became entranced by the White Cat who magically helped him. Having succeeded in the first two tasks, the third task was for the sons to return with the most beautiful princess. The young prince could not believe the White Cat when she instructed him to cut off her head and tail. Reluctantly he obeyed, and instantly she was transformed into a radiant princess, breaking the fairies spell. Such woman/cat transformations worked by love have a long pedigree.

The positive portrayal of the cat in pantomime was part of the improvement in its image wrought by the literary tradition of writers and actors. The cat had powerful friends in the 'PR' world, and after centuries of persecution it certainly needed them.

THE BIRTH OF THE CAT FANCY

You may loathe cat shows or you may love them, but all cat lovers owe them a debt of gratitude for they transformed the attitude of the Western world towards cats. They were still stigmatised by the taint of witchcraft, which had left residual antipathy and antagonism towards cats.

Harrison Weir was responsible for their transformation into their position today as a supremely popular pet.

THE 'FATHER OF ENGLISH PANTOMIME'

John Rich (1692–1761) began pantomime with his adaptation of Harlequinade. At a time in Britain when only the most daring would admit to liking a cat, Rich was one of the most noted lovers of cats, with twenty-seven of his own! The *Dublin Review* noted that an actress found him with 'one licking the buttered toast on his breakfast plate, another was engaged in driking the cream for his tea, two cats lay on one knee, one was asleep on his shoulder, and another sat demurely on his head'!

The involvement of key people like Rich who doted on cats, and the arrival of pantomime's origins through lands that at the time were more pro-cat than Britain, led to the selection of positive cat-magic storylines from folktales to overlay Harlequinade. They helped to change the entrenched British ideas about the cat.

35

Words like 'soft' and 'sensuous' are used when describing both feminine and feline characteristics

Opposite: Matching elegance: a Somali cat and its owner at a cat show

Renowned as the Father of the Cat Fancy, Weir was also documentor of its birth both in his justifiably famous book *Our Cats And All About Them* and as an illustrator of cats. He had a passion for cats and was distressed at their continual shabby treatment, so resolved to change it.

Weir had a lifetime's experience as a senior judge at pigeon and poultry shows, and by bringing together his love of cats and the practical knowledge of the mechanics of showing, he established the basis of exhibiting cats. He convinced his friend Mr Wilkinson,

manager of the Crystal Palace, that it was 'a thing to be done'. Weir recalled the events that followed:

'In a few days I presented my scheme in full working order: the schedule of prizes, the price of entry, the number of classes, and the points by which they would be judged, the number of prizes in each class, their amount, the different varieties of colour, form, size and sex for which they were to be given. I also made a drawing of the head of a cat to be printed black on yellow paper for a posting bill.'

Cats had been exhibited as part of

agricultural shows from time to time, and cats were on display at St Giles Fair, Winchester in the sixteenth century. In that era, however, it would have been a very different exhibit. Weir's action was a watershed, for by putting 160 cats on show in the glass Crystal Palace they were seen by a very wide audience. His first Crystal Palace show was held on 13 July 1871. The ruse worked spectacularly, cat shows became immensely popular and the cat was transformed into a much loved animal.

Weir's standards of 'Points of Excellence' have been followed in essence over the years. A National Cat Club was established in 1887 as a registering body for recording cats' pedigrees. In 1910 this function was taken over by the Governing Council of the Cat Fancy (GCCF) which was formed for the purpose.

In America, pride in the rugged farm cats of Maine had also led to local shows of the breed, possibly before the first national multi-breed show launched an American cat fancy. That began when James Hyde, who had experience of holding horse shows for many years at Madison Square Garden in New York, attended a Crystal Palace Cat Show while he was staying in London. On 5 May 1895, he opened the first New York show. He arranged for some eight British star cats to be shipped to New York to appear at the show to display the standards that had been achieved by showing over the quarter of a century of experience in Britain. Sadly, the temperature rose to 95°F and some of

the British cats died, but others, like Topaz and The Banshee, went on to form notable dynasties.

In Britain, the GCCF has been the sole arbiter of cat showing and registration for most of this century. In 1983 it was joined by the alternative organisation the Cat Association of Great Britain. In contrast, North America, due to its size, has a history of a multiplicity of registration bodies among which is the world's largest, the Cat Fanciers Association (CFA). European countries have their own registration bodies, but these are usually linked to the Fédération Internationale Féline (FIFe).

Appropriately, the oldest national club holds Britain's National Cat Club Show. That is held in December each year at Olympia (under GCCF regulations), it remains the largest event in the world with over 2,000 cats on show.

THE CAT FANCY – A WOMAN'S DOMAIN

Although growing numbers of men participate in the cat fancy, it is till numerically dominated by women, and this was even more the case at its infancy. Although the cat fancy was started by a man, Harrison Weir, with the Crystal Palace Cat Show in 1871, his founding of the National Cat Club in 1887 was followed by that of the Cat Club in 1898 by Lady Marcus Beresford. At the turn of the century 80 per cent of those organisations' committee members were women. The ancient association of cats and women still continues.

Still living wild

Above: Two of my Fitzroy Square colony (Diag and Double-eye Flick) looking out from their usual basement-steps sleeping area, protected by railings

Top: Cars provide invaluable shelter for urban feral cats; from cold and rain in London, and from the full heat of the sun in Spain, as seen here

Opposite: Sketches of six of the Fitzroy Square feral cats, showing the characteristically individual black-and-white face markings that aided identification. Their names were mainly derived from these

The cat was one of the last animals to be domesticated, thousands of years later than the dog, and it readily reverts to living wild. The study of gone-wild or feral cats has taken up much of my life since I started field observations on cats in the mid-1970s.

Most people had only considered cats as pets, so their very familiarity meant that the wild life of domestic cats had not been thought worth studying, even by biologists. Urban ecology was in its infancy; biologists had preferred looking at 'proper wildlife' in 'real countryside'. Research work on urban foxes led to statements that 'foxes are the main predators in towns'. However, I realised that house cats alone were far more numerous, and that stray and feral cats

would have an impact – but what? At that time, twenty years ago, it was thought that groups of cats were only 'loose assemblages' with no social structure.

While I kept watch on a number of feral groups throughout London, my main study colony of cats was in Fitzroy Square, W1. The site was ideal for the cats, with protective railings, a large mature garden with good cover, yet also with open grass and sunning spots. The cats had a number of dedicated feeders. one in particular, Mary Wyatt, spent most of her weekly money on the cats.

It soon became clear that this was no random, varying aggregation of cats with different animals wandering in and out. On the contrary, I observed that the same animals held the site day after day, and, as the study went on, year after year. The group were all black-and-white animals – just like T.S. Eliot's 'Jellicle Cats'. They had certainly been in continuous residence for twenty years then, and in all probability for significantly longer.

The cats were nearly all related to each other. When I started the study, a mainly white matriarch ruled the roost, although not by overt aggression. I spent night after night watching the cats, carefully monitoring their movements. As they were all black-and-white, I found quick sketches of their face patterns invaluable. I named them after their face patterns, which helped recognition. Sketching cats makes you look at them in a different way than when taking photographs.

When I began working with feral cats, it was standardly assumed that life for them was nasty, brutish – and short. On the contrary, however, I discovered that most feral cats were fit and robust. To check this I weighed the Fitzroy Square cats, plus those from seven other colonies and found their weights were comparable to house cats. A conflicting myth to the undernourished cat was that they became monster-sized living feral, so I also measured them. Again the similarities with house cats were more noticeable than the differences.

During the 1970s I was one of a tiny handful of researchers throughout the world investigating feral cat behaviour and the size of range they needed. The results of my research saw the light of day in scientific meetings and papers, and in my first book *The Wild Life Of The Domestic Cat* in the early 1980s. It was a fascinating time, for the ranges of the urban-living London feral cats I observed were smaller than those of the dockyard feral groups studied by Jane Dards in Portsmouth, and much smaller than those farmyard cats monitored by David MacDonald and Peter Apps. Yet, remarkably, the pattern was the same; toms ranged wider than queens, using about ten times the area, and the groups clumped around a source of food.

I realised that greater amounts of food resulted in more cats, for with more food available the cats do not

have to travel so widely to find it and therefore their ranges are smaller. I was able to make a straight line graph showing the relationship between the density at which the cats live and the size of their ranges. As time has gone on, work of other researchers around the world has supported my conclusions, notably in 1988 when Olaf Liberg and Mikael Sandell published a similar straight line graph using figures from studies of feral cats around the world.

Until we started to log the cats' movements, no one really had any idea of how the cats' social system of land use worked. Similar studies carried out on both Tigers and Forest Wildcats has shown a cat family pattern of males holding significantly larger ranges than females. However, there did seem to be a difference, for while the ranges of female Tigers and Forest Wildcats do not overlap much, the English feral cat studies showed the cats clumped around a core area that included a major food source. In Fitzroy Square it was where the feeder regularly put down food.

There are social advantages from group living, and the higher density of urban and farm feral cats, compared to wild species of cat, mean it can readily occur, but is it just due to density? The clumping of ranges seems to reflect the pockets of local abundance of food. Queens that have to catch all their food rather than scavenge from a dump or obtain food from a feeder, like those living on uninhabited islands, do not usually overlap their ranges. Yet one of

Protective Queen (female)

Diag (male)

Double-eye Flick (female)

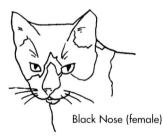

Black Nose (female)

Eyebrows (female)

Flat Cap (female)

The real Angora living in the unchanged, original hillside village of Angora, from which Ankara has developed. Despite its long hair this cat is a working yard cat, for its coat does not tangle

Opposite: A feral cat feeder and some of her cats in Venice. She is a member of the Anglo-Venetian cat welfare society 'Dingo'

the undoubted stars of my BBC *Cats* series was a ginger-and-white feral tom living on the Hebridean island of North Uist. Although he hunted to survive, his range did overlap with other feral cats around an abundant food source, a large rabbit warren.

So, our small Domestic Cats, when living feral, have a flexibility over their ranges that allows them to tolerate other cats if there is a good local source of food. Yet, at the same time, they have the same basic pattern as the rest of the cat family. It is not possible to study the ranges of Tigers or Jaguars over such a range of food availability as it is for the domestic cat living feral, so the studies on our wild-living Domestic Cats have shed light on the range requirements of their larger cousins.

Around the World

I have studied cats living feral around the world, and found that they live remarkably similar lives despite their radically different climates and countries. In rural areas, food availability is insufficient to sustain large numbers of cats, so densities are low. There is more food to scavenge in towns, and in that, feral cats are reliving the pre-domestication period of the species. A significant part of the cat population is always in a feral state. In the ancient bazaars of Egypt there will be some feral cats whose lineage contains few ancestors that have known domestic life. Although that may seem remarkable, it is almost equally true in Britain. Keeping cats as pets in signicant numbers is so recent, even in cat-obsessed Britain; it dates from the invention of cat shows in the nineteenth century, and the rise of suburban living.

In Britain, pet cats claim most of the country's extensive suburban landscape. To obtain food our house cats only have to stagger from their snoozing spots to their food bowls. Consequently, feral cats cannot achieve the same density of numbers as nearby house cats, and require larger areas in which to scavenge for food. The ferals live in the gaps left by the house cats in the suburban landscape, such as hospital grounds, factories and squares.

In many countries house cats are less common than in Britain, and lead wilder lives, at lower densities, like farm cats. In reality the nearer a rural economy approaches subsistence agriculture, the more the cat remains a working animal. In a village in the Spanish Pyrenees, I recall talking to a lady about her cats. The main room of her house was her kitchen in which chickens scratched around the floor. Her cats were rodent controllers of wild temperament. She threw them chunks of meat, over which they growled at each other. I have met the same situation in the hill villages of Turkey and the jungle villages of Nepal and many parts of the world, although the cats usually show less aggression.

Feral Cat Colony Control

The control of feral cats is an emotive issue. In Britain, as in many countries, if an animal was regarded as a 'pest' or 'problem', the automatic response was to call in someone to kill it – albeit humanely.

A reliable food source: a market's permanently sited waste skip in Istanbul has drawn a number of scavenging feral cats for many years; inevitably the cats in the group are now related

In rural areas, gamekeepers on estates despatched large numbers of cats by shooting and by the barbarous and indiscriminate use of wire-noose traps and other traps. A proportion of the cats would be feral, but many were farm or house cats. Fortunately as gamekeepers became an endangered species themselves this practice waned.

Key places where feral colonies were thought to be a problem were large factory sites or hospitals, usually with sizeable grounds. Hospital administrators thought it 'self evident' that all the cats had to be killed. However, this approach usually failed for eradication is rarely total, and cats moved in from other areas. A controlled, stable colony prevents this 'vacuum effect', as I dubbed it. If an area was able to support a population of cats and they are eradicated, then another group can fill its space. However, devoid of the family linkages that existed within the former group, the new cats will breed and be a less stable group.

Although a 'total wipe-out' often appeals to authorities due to its apparent simplicity, it normally fails as it does not allow for biological reality, and worse, is often counter-productive. However, if a colony is neutered and returned to its area, it will continue to hold the location and keep out other cats by its presence. The group's population will gradually decline over a few years. During their brief time in captivity veterinary inspection and treatment can ensure the good health of the group. In contrast if they are removed from the site, other cats would move in, and being of assorted origin would be potentially less healthy.

Over twenty-five years ago Celia Hammond, who abandoned her career as a top catwalk model and adopted the path of feral cat carer, became dissatisfied with the ineffective trap and kill approach and began to put neutered cats back on sites. When I began monitoring my Fitzroy Square cats, before and after neutering, I believe I was the first biologist to assess the effects on a feral cat group. The cats were neutered by the Cats' Protection League, and for many years after I found that the reducing population continued to hold their area.

Paul Rees studied a hospital feral population, while Tom Kristensen of Kattens Vaern was involved in a Danish neutering and return scheme. In 1977 I was asked by Celia Hammond and others to launch the Cat Action Trust which has gone on to neuter many thousands of cats all across Britain. Peter Neville monitored a site in Regent's Park with Jenny Remfry for the Universities Federation for Animal Welfare (UFAW) after Jenny had first tried and then discontinued the possible alternative of putting feral cats on the 'pill'.

Each of these neuter-and-return studies had similar positive findings. Subsequently the practice of neutering and returning feral cats back to their site has been applied widely by organisations, including the World Society for the Protection of Animals in

Island dockside cats in Greece. Their food is seasonal, arriving with each tourist boat in a summer glut

Feral cat carer feeding a neutered feral cat group in Florida. One of the first neuter-and-return projects in the USA

Europe. In 1990, Werner Passanini, with David MacDonald of Oxford University, surveyed the effectiveness of neutering groups for the UFAW and concluded that it is the most effective way of managing feral cats.

As the neutering of colonies became more common in London it brought down overall numbers. Whereas when I began to observe the Fitzroy Square colony, occasionally other cats attempted to enter the group, as years went by this became less as other cat groups in the area were also neutered. As a control system it is able to manage broad areas.

Despite this successful long track record in Britain, the technique became the centre of a heated debate in the USA in 1993. During the early 1990s over 400,000 cats a year were being destroyed in California's cat shelters and welfare workers were looking for a way to reduce this. America's National Audubon Society saw feral cats as a threat to wild birds and co-sponsored the AB 302 Feline Fix Bill, together with some cat welfare organisations, in an attempt to have it introduced into the California State Legislature. The bill proposed to make feral cat carers the legal owners of the cats and insisted on the neutering of all the carer's group

within thirty days, and with fines for non-compliance. The spectre of elderly cat feeders being 'thrown into the slammer' became an alarming possibility. The situation was not just one of 'cat people v bird people' but also 'cat people v cat people'!

Over forty state and national animal groups formed Californians Against AB 302 to have the bill withdrawn. They declared that

'A feral cat care programme aimed at spaying/neutering, vaccinating and feeding represents the grass roots animal control needed to get the problem of cat overpopulation under control. We will no more support a bill that criminalises this work, threatens to hasten the impoundment and killing of cats, or threatens to fine those who do most work to stem the causes and symptoms of California's overpopulation of cats, than we will one that outright targets feral cats for impoundment.'

The National Audubon Society's Californian Legislative Director was quoted in the press to the effect that his position on AB 302 was one of opposition to 'any programme which purports to "trap, spay or neuter or return" cats to the wild'.

The lack of understanding of the effectiveness of the neuter and return technique mirrored that of Britain twenty years before, and the approach of the co-sponsors of the proposed bill was as if the system was new and untried. Yet in the USA itself, Alley Cat Allies (one of the signatories of the anti-AB 302) had been following the British experience and implementing the neuter-and-return

method. They were not alone, for they surveyed a thousand people working with feral cats across North America; 80 per cent replied and of these 91 per cent were using neuter-and-return method.

To find out the experiences of the method outside the USA, and to find a consensus among the cat welfare charities, in 1994 the Doris Day Foundation brought together a meeting of key organisations in Washington DC and invited myself and Jenny Remfry to address it. It was alarming to find that some charity organisers thought it better to put animals down to 'save' them from suffering which they *might* experience at some time. Some assumed that cats living feral *must* be leading a terrible life, regardless of evidence to the contrary.

In a huge cat owning society opinions are bound to vary, but discussing differences is leading to a better understanding of not only the neuter-and-return method, but also of the feral-living cat. While cats are no longer made divinities or demons, attitudes to cats around the world are in a state of flux, and to some in America today the cat is a demon hunter.

Feral cats are great survivors and always manage to find secure shelter. Pipes are quite commonly used

Breeds and genetics

A champion British Blue, the epitome of the full-jowled male British Shorthair show-cat

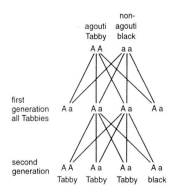

Mendel's 3:1 ratio from the mating of agouti and non-agouti cats (such as Tabby and black cats)

UNDERSTANDABLE CAT GENETICS

In reality cat genetics are not simple, but in outline they follow a straightforward pattern – even if it is full of exceptions!

When Harrison Weir caused human manipulation of cat breeding to come into proper existence by inventing cat shows in the nineteenth century, he argued that: 'Why should not the cat that sits in front of us before the fire be an object of interest, and be selected for its colour, markings and form?' At the time, there was no understanding of the mechanism of genetics and his rules for breeding were the soundest advice available:

'In the first place, the fancier must thoroughly make up his mind as to the variety of form, colour, association of colours or markings by which he wishes to produce, if possible, perfection; and having done so, he must provide himself with such stock as, on being mated, are likely to bring such progeny as will enable him in due time to attain the end he has in view. He has to build up a family with certain points and qualities before he can actually embark in the real process of accurately matching and crossing so as to produce the results...he is hoping to gain eventually.'

It was only at the beginning of this century that biologists stumbled on the obscurely published work of Gregor Mendel who discovered the principles of heredity in the 1860s. His breakthrough was to identify that characteristics are inherited by means of units that we now call 'genes', and he also discovered the working ratios in which these are passed on in successive generations.

We know now that genes are sections of chromosomes, made of DNA. Each chromosome is a chain of genes, and every body cell of the cat has nineteen pairs. The special type of cell division for the sex cells makes each one have only nineteen chromosomes. During the division that takes place to form individual sex cells, parts of one half of the divided chromosome may swap with parts of the other which randomises the genetic components. When adult sex cells fuse together on fertilisation, they re-form a full complement of thirty-eight chromosomes.

Each new kitten inherits a gene for

each characteristic from both of its parents. Early in the history of domestication, all coat colours and patterns were the same – mackerel tabby – so the coat genes were identical. The replication of the DNA during cell division for the sex cells produces an identical copy. However, mistakes can occur and be copied. This is a mutation. The original gene and its flawed image are termed 'alleles'. Although an incredibly rare event, it is nonetheless the origin of the range of coat colours, types of coat and other genetically changed characteristics in Domestic Cats today.

Most mutations are harmful and do not survive – but some do. The earliest coat mutation was the appearance of the black cat. This is also seen in thirteen species of wild cat and is called melanism.

Early cat breeders like Harrison Weir knew that if you crossed two apparently identical tabby cats a black kitten or another solid colour kitten could be among the litter of tabby kits. Usually, one coat colour is dominant over another, which is recessive. Tabby is dominant over black. To achieve consistent genetic results, Weir advocated that a 'pure' family of tabbies would always breed 'true', and similarly a line of black cats would as well. If one of these tabbies and one of these black cats were mated together their offspring would all be genetically dominant tabby, but if two cats from this generation were mated together there would be both tabby and black kittens, Mendel's ratio of 3:1.

The situation in the cat is complicated because the tabby does not really disappear. Genetically all Domestic Cats remain tabbies despite the evidence of your own eyes. Even in a solid-coloured cat you can still see the bands of the tabby markings in a particular light. The tabby coat is an alternate pattern of solid colour and flecking (called agouti, see pp23–4). An individual hair of a tabby is banded with yellow which gives the agouti section, and in the solid colour part of the coat this hair banding is absent and is called non-agouti. Consequently the black and tabby cross is, genetically, really one of agouti and non-agouti.

When the alleles are the same in a pair the cat is called homozygous for that characteristic; when it is mixed it is termed heterozygous.

Some colours occur as a result of additional genes modifying others. Black is turned to chocolate by a further recessive gene. Black has an additional gene to be a full dense 'colour', but an allele of that will turn black to grey (called blue). This 'dilution' gene changes the distribution of pigment in the hairs, so some incident light is reflected. Brown is diluted by the gene to lilac in the same way.

These types of straightforward results statistically occur readily in breeding with most of the

A three-year-old female Somali with its natural shaggy coat; it has the appearance of an historic breed, yet has been developed over the last thirty years

other coat colours that have historically mutated subsequent to black.

The Exceptional Ginger Gene

Some inheritance works by a more idiosyncratic manner, and that of the ginger or orange colour does in particular. It is usually assumed that genes are inherited randomly, but some genes are on the same chromosome and are inherited together, linked. One pair of the chromosomes dictate a cat's sex - females have a matched pair of chromosomes designated as XX, while males have an X and a Y. The rest of the chromosomes occur in matched pairs, but the Y is smaller than the X. While every ovum has an X chromosome, a sperm can have either an X or a Y. Consequently any mutant allele, such as that for orange, carried on a sex chromosome, will be 'sex-linked'.

The Y chromosome does not have a site for an orange gene, so a male can only transmit the gene on the X chromosome, and can only be orange (O) or non-orange (o). In contrast, a female, with two sites can be orange (OO), tortoiseshell (Oo) or non-orange (oo).

There is nothing straightforward about the ginger gene, so although in a tortoiseshell female with Oo there are two alleles, there is an attempt to compensate for the male having only one. One of these alleles in each of the multiplying number of skin cells in the early embryo is ignored. This happens randomly so a mosaic of orange and non-orange skin cells develop. The tortoiseshell queen develops this way, but if she also carries the gene for piebald spotting and so has white as well, then the tortoiseshell pattern becomes more distinct; it is called Calico in the USA.

On infinitely rare instances a cat has two X chromosomes and a Y chromosome, but these 'males' are usually sterile. Some variations can be caused genetically by modifying groups of polygenes. Rufus polygenes enhance the degree of richness of colour.

Incomplete Dominance

While most genetics work on the all-or-nothing basis of dominance or recessive, the notable exception is 'incomplete dominance' in which both genes function but neither is dominant. The prime examples are the pointed cats of South East Asia. When a Seal-point Siamese is crossed with a traditional Burmese, a midway colour appears of a less full colour form of Burmese, but with colouring to the points. This is the Thai Copper and the West's Tonkinese.

THE BREEDS

The simplest view of 'What is a breed?' is what is recognised as a breed by a registering body. Consequently, a breed in Britain may be different from what is considered a breed in the USA. However, from the biological perspective, criteria for some modern breeds are trivial. In contrast, historic breeds have formed as a result of geographic and therefore genetic separation, so although their

A female tricoloured Japanese Bobtail, orange, black and white, the *mi-ke* cat. These cats were often depicted in nineteenth-century Japanese woodcuts

development was aided by humanity, natural selection has largely been involved. These ancient types of cat are the most authentic of breeds and were formed long before the advent of the cat fancy.

SHORTHAIRS

The British Shorthair should really be the robust 'moggie' whose story was related in the foregoing chapters, but during World War II, breeding was drastically reduced. Then when pedigree numbers were low in Britain, they were bred for no good reason to foreign body shorthairs, and then mated to Persians to correct the build. The result was a different cat. It is an attractive animal, but it does not represent its historic lineage. Its beak is inherited from the Persian.

The same happened in America, but the CFA called the new cat the Exotic Shorthair, while retaining the name American Shorthair for what purports to be the all-American mog. Yet in the early days of the American cat fancy, British cats were mated with American cats.

The loss of the basic mog from British showing would have saddened Harrison Weir who said: 'a high-class, short-haired cat is one of the most perfect animals ever created'.

The European Shorthair had the same traditional build, yet there could be distinctions in breed lines. The blue cats illustrate the differences between history and reality. Tradition informs us that Carthusian monks bred a distinct blue cat called the Chartreux. However, to

keep the Chartreux in existence cats were mated with British Blues and Persians, which eroded any distinctiveness between the cats, so both are judged on the same standard. Authenticity fell before the face of expediency.

Fortunately in the early 1980s in Scandinavia a group of breeders began the reinstitution of the selected mog as a showcat again, which led to the recognition of this European Shorthair

A colourful Calico cat, tortoiseshell and white, making an impression in Monet's garden at Giverney, France

Top: A traditional build Brown Tabby Persian; a homesteader's cat in Australia

Above: The true Van Cat in Van, Eastern Anatolia, Turkey. For the people of Van the features that make their cat special are the odd-coloured eyes and a snow-white silky coat. The orange-marked cats recognised in the West do occur in Van, but are not highly regarded

by the Fédération Internationale Feline (FiFe). Housecats have been shown in the novice class for European Shorthair at Cat Association of Britain shows as the Association is a member of FiFe.

LONGHAIRS

My earliest recollection of a cat goes back to my childhood days and to my grandparents' lovely old Brown Tabby Persian. It spent much of its time asleep

in the afternoon sun on the wooden lid of a massive waterbutt at the back of their house. It was a fine cat of a type popular forty years ago with a fine, straight unmessed-about nose and an old style coat that did not need brushing.

Frances Simpson, the influential show judge, writing in 1903, shared some of my childhood sentiments:

'I cannot explain it, but certain it is that of all the feline race the warmest corner of my heart has always been kept for the brown tabbies (Persians). There is something so comfortable and homely about these dear brownies – they seem to have more intelligent and expressive countenances than any other cats, and I am firmly of the opinion that no Persian cats are so healthy and strong as brown tabbies'.

Angoras

Weir recalled that around 1850 most people referred to longhair cats as 'French cats, as they were mostly imported from Paris and more particularly the white, which were the fashion'. These were the Angoras that had been the pampered pets of the Ottoman Empire and had travelled to France to become the darlings of the seventeenth-century French court. In Britain in 1889, Weir was still able to distinguish between the Angora, the Persian and the Russian, and these are distinctions that we can draw again today between the Angora, Persians and Northern Longhairs including the Russian.

However, by 1903, Frances Simpson

could find 'hardly any difference between Angoras and Persians', and the distinctions were so fine 'that I must be pardoned if I ignore the class of cat commonly called Angora, which seems gradually to have disappeared from our midst'. This happened due to indiscriminate breeding between all longhairs regardless of origin. Fortunately, attention to historic and geographic integrity of breeds has again become recognised as important.

From Istanbul to Tehran, longhair cats can be found with a slim, warm-climate build. The genetic mutation selection for these longhairs has been favoured by conditions that can inflict severe winters yet scorching summers. The Anatolian Angoras and Vans have an elegant long-faced appearance. The reappearance of the true Angora in the show world since the 1950s and 1960s has once again begun an appreciation in Europe and America of these most charming of cats.

Modern Persians

Persians of the form and build of my childhood companion shared something of their ancestors' build. However, breeding developed in Britain and Europe to foreshorten the animal's nose to what its devotees view as a more appealing face. To its detractors the flattened face gives a more pugnacious appearance. The adherents of the modern Persian Longhairs are not concerned by the criticism that by imposing a flatter 'baby' face and a fuller coat that needs daily brushing, they are perpetuating a dependent toy, but they believe that for the sake of

appearance the extra grooming is a requirement they are happy to meet.

In temperament, most Persians are usually docile and gentle. Their appearance has led to their long domination of the showbench.

Silver Tabbies, Smokes and Chinchillas – An Inhibitor Sequence

The Longhaired cat exists in an array of types but some stunning coats began to be bred at the end of the last century. The Smoke Persian was granted its own class in 1893 and the Chinchilla followed in 1894. These emerged from breeding Silver Tabbies and form a series around an inhibitor gene.

In the Silver Tabby, the yellow banding is suppressed by the gene more than the dark lines, so the coat is black tabby on white. The Chinchilla has the most remarkable of marked coats, for it is as if the mist from the ghost of a tabby has kissed the tips of the hairs. Its delicate shading is due to the inhibitor gene suppressing colour in the hair, except at the point where the hair first grows. The difference between the Silver Tabby and the Chinchilla is just one of degree effected by polygenes.

An eighteen-month-old, female Blue Colourpoint Longhair, with the coat and build of the modern Persian

The Smoke is caused in the same way as the Chinchilla, but as it contains the non-agouti gene its tipping is more noticeable and even.

In Cameo Longhairs, orange is swapped for black.

Colourpoint Longhairs – East Meets West

One of the largest group of current showbench cats are the Colourpoint Longhairs, called Himalayans in the USA. These cats are a blend of East and West, for they inherit points of colour from the Siamese in the body of a modern Persian. While breeding research towards such a cat began early this century, in both Britain and America it was developed some forty years ago.

To keep the predominantly Persian appearance, breeders usually cross Colourpoint Longhairs with Persians rather than with another Colourpoint Longhair. The resultant solid-colour Longhairs are then crossed with a Colourpoint Longhair as the colourpoint gene is recessive, and their litter yields both Colourpoint Longhairs and solid-colour Longhairs. The result has been the Persian reinforcement that led to the cat being a coat-colour class of Persian rather than a 50:50 blend and a separate breed.

OTHER LONGHAIRS

The Balinese

The Balinese is a longhaired mutation of the Siamese. It has the body of the Siamese with flowing silky hair. It retains the outgoing, talkative nature of the Siamese, but shares with the Angora a near tangle-free coat and still retains an elegant, long nose. In recent years, the conformation of the Balinese has pursued a svelte form.

The Birman

The Birman shares the Siamese points and the

The newer type Balinese sitting alongside the older type lying down build of the earlier-bodied Siamese breed type. Its white feet are the distinctive feature. There is a belief that the white feet derive from the cats from Lao-Tsun Temple in Burma. A folklore story suggests a supernatural origin for the first cat's white feet from contact with its dying companion, the priest Mun-Ha.

The Somali

As the Balinese is to the Siamese, so the Somali is to the Abyssinian. It is one of the most striking of cats, and due to its ticked coat, has 'the soul of the wild'. The Somali probably originated as a mutation, but it has been argued that the recessive long hair gene stems from early outcrossings to longhairs.

NORTHERN LONGHAIRS

For those who prefer the older, rugged style of Longhair, the growing interest in the various Northern Longhairs has been met with enthusiasm. The massive presence of the Maine Coon Cat from New England has a growing following in the Old World. In the USA they are most popular.

The similar Norwegian Forest Cat is equally robust with a similar homestead background.

ORIENTAL AND FOREIGN SHORTHAIRS

Previously in Britain, under GCCF ruling, the term 'foreign' was applied to all shorthairs except British-type, and 'oriental' was used for patterned coats. Since 1991, Siamese-derived cats are dubbed 'oriental' and all others are

classified as 'foreign'. The term 'foreign' is still generic for conformation, and consequently is applied to Rex cats, despite the lack of geographical reality.

The cats of East Asia are distinct from those of European ancestry. Among street cats in East Asia one of the most noticeable genetic modifications is the kinky tail, which occurs in two out of three cats. Its abundance suggests that cats were introduced to the tip of the Malay Peninsula by Arab and Indian seafarers centuries before any European

Top: The magnificent Maine Coon lived as a semi-feral farm cat in New England, where its full coat protected it from the harsh snow-bound winters of Maine

Above: The Norwegian Forest Cat shares the Maine's weatherproof warm coat. The Danish *Racekatte*, the Swedish *Rugkatt* and the Russian Longhair are the geographic equivalents of the Norwegian *Skaukatt*

53

exploration. Only a few cats arrived initially, and a malformation in the tail of a cat in this early population became widespread. It does not carry the dangers of the Manx gene.

The extreme form of kinky tail which reduces the tail to a rabbit-like bobtail is the basis of the Japanese Bobtail, which has a long history. The slender build of the original cats from the Mediterranean ancestry has been retained in the warm climate of South East Asia.

In Thailand centuries ago, the *Cat Book Poems* were written down under royal instruction, which document the Siamese, Korat and Copper and reveal the antiquity of these breeds.

The Siamese
The Siamese was the consort of the monarchs of Siam, and was presented to certain visiting dignitaries in the nineteenth century. Its remarkable appearance with dark points at the first cat show in 1871 caused a sensation and was a major factor in the show's success. Initially Siamese in the West were only imported in the dark

sealpoint, but gradually the hidden inheritance of recessive and colour dilution genes gave the blue, chocolate and lilac 'classic' Siamese colours.

Among the temple cats of Thailand there are other colourpoints that have arisen by natural crossing. Breeders in the West have extended the Siamese inheritance into Longhairs, making it the main coat determinant (after the underlying tabby) in the show-cat world.

The Burmese

The Burmese was believed to have been a breeder's invention in the West, as it appeared when a US Navy psychiatrist took home a brown cat from Rangoon in 1930 called Wong Mau; it showed some point darkening.

It was crossed with a Siamese, and then the kittens were crossed back to Wong Mau. In the litter along with Siamese offspring were also brown cats and a rich dark chocolate cat which was designated Burmese.

Brown cats were termed Thong Daeng or Coppers in the *Cat Book Poems*. I found both Coppers and full Burmese feral in Thailand. Breeders in the West of the Tonkinese – the remade Copper of the East – were delighted to learn that the breed had a history and could be found today.

Burmese are currently a 'smart' cat for the young owner who would not be seen dead with a fluffy modern Persian, but who feels that he or she needs to be 'upmarket' of a moggie!

Opposite top: Professor Edward Rose and his wife Malee at their breeding cattery in Chiang Mai, Thailand. He is holding two young Burmese, the progeny of the robust Copper stud tom called Woot held by Malee. This is part of a programme to broaden the Burmese genetic base in the West with genes from the original area

Opposite below: Illustrations in a Thai manuscript *Cat Book Poems*. The upper cat is a depiction of the Siamese, or 'Vichien Mat'

The Siamese, 'The Royal Cat of Siam' in its authentic original build; in Thailand

Top: The Korat is considered a 'good luck' cat in Thailand. Its late acceptance in the West has protected it from the build changes inflicted on the Siamese and Burmese

Above: A Russian Blue. Their distinctiveness was threatened when before World War I Russians were judged to the same standard as British Blues

Opposite top: A Peke-faced Persian in the USA

Opposite below: A young female chocolate Oriental Spotted Tabby

Unfortunately fashion has moved the Burmese towards a slighter build in Britain, while the original form has been retained in the USA.

The Korat v the Russian Blue!

The blue cat from the Korat province of East Thailand is one of the true gems of the cat world. Originally the farmers' cat of Thailand, it was termed 'Si-Sawat' in the *Cat Book Poems*, which noted that its 'eyes shine like dewdrops'. It is a gentle but lively cat, and popular with breeders today in Thailand.

Although shown in Britain at the end of the last century, the Korat was ignored as the Russian Blue was already established. The Russian Blue's arrival was attributed to sailors bringing it from Archangel.

Even when the Korat became

established thirty years ago in America, and was then re-introduced to Britain, there was again resistance due to similarities to the Russian Blue. Confusion was not aided by a post-World War II outcrossing of Russian Blues with Siamese blue points to change them to a more foreign type. However during the 1960s Russian Blue breeders did a 180° turn in an effort to regain the original conformation. However, the Korat eventually won acceptance.

The Oriental Shorthair Selfs

In the early 1950s, at the time that the brown Burmese were being introduced into Britain from the USA, breeders in Britain bred a solid brown cat from the chocolate pointed Siamese. This was followed by a foreign-built white, a lilac then a black. At the time it was not realised that Siamese-bodied solid colour cats already existed in Thailand.

Spots and Stripes

Slim, foreign-build tabbies exist in abundance in South East Asia, and gloriously spotted ones in India, and at the birthplace of the species in Egypt and around the Mediterranean. Spotted cats from Egypt gave rise to the Egyptian Mau breed developed in the USA in the 1950s.

Despite this, breeders in Britain pursued a re-creation of the early Egyptian cats starting with a Siamese form tabby. This gave rise to the Oriental Spotted Tabby, which was briefly also called the Mau in Britain. Each of the Siamese-derived cats exhibit the behaviour patterns of the Siamese, but being of the thinner contemporary form are more active than their traditionally shaped relatives.

THE CHANGING CAT

The formation of new breeds has come about this century as a result of crossing and mutation selection. Unlike in cats, mutation selection in dog breeding has retained a wide range of shapes and sizes. Unfortunately the transition from work to showing led to numerous inheritable problems. Many large dog breeds have a tendency to hip displasia. Present day bulldogs are a travesty of their predecessors, as their foreshortened faces make breathing difficult and their disproportionately large heads cause significant problems in giving birth.

In the cat show world as yet the genetic problems are nothing like as widespread. However, there is a clear need to take stock now before such concerns are common among cats.

Almost inevitably, breeding and judging to a written standard takes a breed away from its forebears. The two most notable changes that have affected the largest number of breed cats are to the nose of the Persian (Longhair) and the build of the Siamese. Fortunately for Persians (Longhairs) in Britain, a halt was called to the trend, but in the United States it continued and led to the recognition of the Peke-faced Persian. However, in Britain the accentuated form termed the 'ultra' has caused concern.

The flat-faced Persians have a kitten-shaped face as an adult cat and so have their followers, yet this shape can cause problems for the cat. At a cat show in the USA, I remember in particular one Peke-faced cat which was the centre of its owner's attention and affection. It was therefore very sad to watch as she repeatedly dabbed away the tears that continually fell down its cheeks, before she took it for judging. It was a good natured animal, but to walk the two steps from the centre to the front of its cage to greet

The promotion of a hairless mutation as a breed termed the Sphynx has caused serious controversy

Selection for thinner Siamese cats has led to a change of head profile and skull shape

me left it wheezing. Breathing difficulties and faulty teeth are common in these cats. Even among apparently 'good type' cats at shows, there is a likelihood of running eyes, as the draining ducts are kinked and compressed. On some white show Persians, this can leave a reddish staining that is treated daily with an antibiotic.

I am firmly in favour of the stand taken by the Governing Council of the Cat Fancy which is against the recognition of this breed in Britain. I argue that to choose to inflict distress, to breed to perpetuate disadvantage is wrong, and should be regarded as unacceptable as other causes taken up by animal welfare campaigners today. For me it is an animal rights issue.

In North America, the most popular shorthair showcat, the Siamese, has disastrously collapsed. While twenty years ago the benches of any major show would have been teeming with Siamese, by 1990 they had all but disappeared. When the Siamese first arrived in the West, they were seen as exotic, but they were firm, robust animals. Yet recent fashion dictated they should become more 'foreign' than the real thing. The trend to thinner animals resulted in frailer cats that were hard to breed, vulnerable to illness, and increasingly uneconomic as they became less popular with the public.

Unfortunately, awarding schedule points can push towards unhealthy extremes, and judges need to be aware of not exaggerating trends to the disadvantage of the breed. Part of my *Cats* series that looked at modern breeding trends elicited a particularly strong reaction among many viewers worried by mutation selection for weirdness rather than for good-looking cats. Although dogs have undergone drastic changes in appearance as a result of breeding, in cats this is unprecedented. Domestic cats have looked after their own breeding for 3,500 years and function was paramount for survival. Yet in a matter of a few decades we are now causing serious problems.

Novelty or Alarming Trends

I prefer a cat to look like a cat. Breeding to enhance and not disadvantage the animal is a fair guideline. Unfortunately reasoning is not always straightforward. The longhair kittens appearing in Siamese litters before 1950 were put down as freaks, yet the Balinese line that developed from later identical kittens is delightful. Since 1950, other conditions previously considered genetic abnormalities have been taken up as a basis for a breed.

In the 1930s Rex cats were viewed as

freaks. Yet forty years later they had gained acceptance. The next step, a move away from the normal coat, caused dismay to many people. In 1966 the naked Sphynx cat was born in Canada, and was developed into a line in North America. There is something distasteful about a judge looking for 'nice wrinkling of the skin' at a show.

Ears, too, have been changed. In Scotland, in 1961, a cat was found with flattened ears. The line that developed was disallowed registration by the GCCF in Britain, but was acknowledged by the CFA. Consequently, America became the centre for 'Scottish Fold' breeding. Breeding must always be to non-fold cats, for the homozygote has gross thickening of cartilage in the feet and around the joints which prevents walking. In California, in 1981, a cat with ears bent backwards instead of forwards became the 'American Curl'.

How far will people go in the pursuit of the bizarre regardless of the animal's welfare? Potentially there is a 'four-eared' breed as cats have been born with an extra pair of ears. However, the head was oddly shaped and the animal

was inactive, which may have been due to damaged brain function. What was once undefendable may now be snapped up, for the hype used to popularise the Rag Doll in the USA was based upon its lethargic inactivity – it was 'ideal for apartments'!

Cats with short legs were reported in the 1930s and 1940s. This is now proposed as a breed in the USA as the mutation has reappeared and been given the name 'Munchkin'. Due to its stumpy legs it cannot jump like a normal cat. This has startled many people, yet the American geneticist Solveig Pflueger says it is caused by the same gene as in the corgi. However, given the problems in reduced-leg dogs that is not reassuring.

Cats were born earlier this century with stunted front legs but normal length back ones, which dramatically affected how they walked. In sitting up the unfortunate animals were described as looking like small kangaroos. What a

Ear movements and positions are a key part of cat-to-cat communication. The American Curl (left) and the Scottish Fold (above) are mutations with permanently distorted ears

59

The American Rag Doll, seen here with its original Californian breeder; it goes limp on being handled, hence the name. Although this behaviour is the characteristic on which the breed was based, breeders in Britain are now seeking to achieve the appearance without the limpness, which has been criticised

This stumpy-legged mutation has been called the Munchkin after *The Wizard of Oz*. The choice of 'cute' names for disadvantaged mutations unfortunately seems to make their being considered as a breed more acceptable

grotesque selling point that could become for a new 'breed'. Will the genetically inherited split-foot and hare-lip become the basis for new breeds? At what point should a line be drawn?

It could be argued that as cat breeding is of minority interest, why worry? Yet as responsible pet ownership has led to around 80 per cent of housecats being neutered, there is an increasing pull of breed cats into the housecat population. While in the hands of responsible breeders problems should be restricted, but as the circle widens,

controls will be harder to implement, and Pandora's Box will be open.

Harrison Weir left his beloved Crystal Palace Cat Shows when he found some owners more interested in prizes than in the welfare of the cat. Judging from his comments when discussing breeding, some of today's changes to the cat would horrify him:

'There are rules which ought to produce certain properties that may be desired, either by foolish fashion, or the production of absolute beauty of form, markings in colours, or other brilliant effects, and which the true fanciers endeavour to obtain. It is to the latter I shall address, rather than to the reproduction of the curious, the inelegant, or the deformed, such as an undesirable number of toes, which are impediments to utility.'

Our judgement is sometimes clouded by considerations that are not in the best interests of the cat. The Manx is the classic case: it is a historically regarded breed, yet a genetic nightmare.

The Changing Cat – Conclusion
Registering bodies worldwide should disallow any physically disadvantaged type of cat proposed for breed status in the future. Those who love the cat should not damage it or promote its damage. It is within the hands of us all to ensure that cats remain the supremely beautiful, healthy, independent animals that they have always been, and that we do not ruin what we love, leaving as our inheritance a genetically damaged and physically crippled animal.

THE HISTORIC CASE – THE MANX

The earliest record of the Manx cat from the Isle of Man, was of those owned by the painter Turner in 1810. It was commonly believed they had developed from cats that had had their tails cut off and had bred tail-less kittens.

As an historic breed, it has an associated folklore, and is part of the national Manx identity. Having spent time on the island with the indigenous cats, I can vouch for their charm and distinctiveness. But if this mutation occurred today, it would be considered too danger-ous to breed.

The Manx gene is dominant, but homozygotes die before birth. Even among heterozygotes, mal-formation ccauses a high rate of still births, and consequently it is a 'lethal' gene. The surviving cats range from having some tail to the 'Rumpy' exhibition form without tail vertebrae. In this state some degree of spina bifida is common, and the pelvis and spine can be fused restricting the cat's gait.

America has become the centre for breeding. Some years ago the Manx Parliament, worried that the cat would disappear from the island, set up a government cat-tery. After ups and downs this was bought out by a private couple who closed it down as they were concerned about the welfare of the cats. Once again breeding on the island has returned to the cats themselves, which is perhaps the best outcome.

Structure
for function

The contrast between the Chinchilla's coat and the black rim of its eyes makes it particularly dramatic in appearance

MAKING SENSE

Cats have specialised sense organs to enable them to exploit their nocturnal, semi-arboreal lifestyle. Nonetheless, many owners attribute additional abilities to their pets. A Cats Protection League survey found that 56 per cent of owners believed that their cats had psychic abilities.

Reflect On Cats' Eyes

The cat as hunter has forward facing eyes. Herbivorous prey species have near all-round vision by having eyes on the side of their heads.

We share a similar eye structure with the cat, but they have specialist adaptations for night life. Cats can see at a light level one-sixth of the amount that we require. Relatively, they have the largest eyes of all domestic animals. Consequently, they have proportionally larger pupils and can let in more light at night. The cat takes in 50 per cent more light than we do. To maximise light intake cats' eyes have a large lens and the cornea is noticeably rounded, giving them a 'glassy-eyed' profile.

This feline nocturnal advantage would be a huge daytime disadvantage without a compensating arrangement in the iris. In most big cats the pupil shuts down in a circular way, but more tightly than ours. The pupil of the Domestic Cat and other small cats narrows to a slit, for it needs extremely fine control to avoid dazzle.

The key feature for which cats' eyes are known is their reflective quality, which lent 'divine powers' in Ancient Egypt, demonic ones in medieval Europe, and emulation in road markers. The reflection is caused by a crystal mirror, the tapetum lucidum. As there are few photons of light in dim conditions, any that missed detection in the retina on arrival are reflected back from the mirror behind the eye to give the retina another chance. The tapetum does not function until kittens reach twelve to fourteen weeks when the need to hunt at night arises.

Cats' eyes have another feature that we do not share – the nictitating membrane, or third eyelid. It is white and often comes as a shock to owners who see only white voids instead of the eyeball. It is perfectly natural and will sometimes cover the pupil when the cat is dozing, allowing light in but losing all detail. As soon as a shadow crosses the eye, the membrane flicks aside and the cat is alert. If something enters the eye the membrane can become inflamed, and it can drift across the eye when your cat is feeling ill. Unexpectedly, as cats do not have the protective

antibacterial enzyme lysozyme in their tears, they blink infrequently, at times only once in five minutes.

Hearing

For the night hunting cat, finely attuned hearing is vital. When your cat hears prey, it is instantly alert with pricked up ears. Even while sitting on your lap its ear cones, or pinna, turn about like radar sweeps. We can hardly move our ears at all, but the cat's pinna have over twenty working muscles which allow the ears to turn sufficiently well to pick up sounds from behind the head. The cat's ability to detect sounds with precision is reduced if it makes

> **COLOUR VISION**
>
> Cats can see colour, but only at a fraction of our ability. This apparent loss has been traded for a great gain in nocturnal vision. At low light levels, we find the landscape drained of colour saturation. In both our eyes and cats' eyes there are light receptors in the retina – cones discriminate colour, and rods are more sensitive at low light levels. Three different types of cones work at different wavelengths to interpret the colour spectrum, similar to the working of a colour TV system. Rods give the equivalent of a black-and-white television picture. With a ratio of twenty-five rods to one cone, in contrast to our four to one, cats' sensitivity in low light levels is massively increased. Behavioural and physiological studies have shown that the little colour cats can see is primarily green, with a minute amount of blue.

Lying down, but alert of eye and ear. At the lower outer edge of the ear the pocket termed the bursa allows flexibility to the ear's movements

The cat's nose is sensitive to temperature as well as scent

dog's prey does not make much high frequency sound, unlike the small mammals and birds hunted by the cat.

Physiological responses have been found at 100kHz, but the realistic upper area of normal hearing above which sensitivity is low is around 35-65kHz. In practical terms, cats have been trained to respond to tones up to 60kHz. Rodents' squeaks are in the 20-50kHz range, which is mainly above the normal human responsive limit of between 15-20kHz, but exactly in the region of good sensitivity of the cat. Although physiological responses for dogs have been found at 60kHz, their realistic upper area for normal hearing is around 15-35kHz.

On the underside of the cat's skull towards the rear corners are two large bulbous areas called the auditory bulla, housing the inner ears. Its formation in the Feloidea is different from that in most mammals. The bulla creates an air space internal to the tympanic membrane, and seems to allow resonances to particular frequencies, and give a greater sensitivity to those frequencies. The cat's disproportionate sensitivity to rustling sounds made by small prey may be due to resonances forming within the cat's ear to those sound frequencies.

Sensing Scents

Inside the cat's nose, separated by a nasal septum, is a labyrinth of bony, plate-like projections, the conchae, which almost fill the space. These are covered by an olfactory mucosa providing a surface of around 20-40sq

directional swivels on the move, so most are made while frozen motionless. The same muscles allow the cat its great range of facial expressions, even to the extent of lying the ears flat.

One of the enigmas of the cat's ear is the bursa. Although this flattened pocket at the outer angle of the pinna seems to have no obvious function, it allows the ear to fold and move, and it may also dampen complex sounds received from behind the head and assist localisation. Cats identify the source of a sound mainly by detecting a difference in the intensity of the sound between the ears. Cats detect sounds at different heights better than dogs, which accords with the cat's semi-arborial hunting.

It comes as a surprise to most people that the dog's ability to hear in the ultra-sonic is less than the cat's. However,

cm – twice the human amount. In the mucosa are olfactory cells that detect the volatile substances that waft in. These cells are only at the top back of the cavity, to which air only penetrates on definite sniffing by the cat rather than just breathing.

Tasting Scent – 'Flehmen'
Below the floor of the nasal cavity are curved cartilage tubes, the vomeronasal organ, or Jacobson's Organ, which connects to small holes behind the upper incisor teeth. This organ enables the cat to 'taste scent' by gaping and adopting a curious trance-like state when sniffing at cat urine or some other scent; the 'Flehmen response'. This German word has no ready counterpart in English; 'grimacing', although the nearest equivalent, hardly conveys the action.

The Jacobson's Organ is only rudimentary in humans, so we cannot appreciate the information it could yield. However, in most animals it seems strongly linked with sexual behaviour. The organ is connected to

A Blotched Tabby giving a characteristic Flehmen response to a scent, its mouth held open

The cat's spiny papillae are most pronounced in the middle of its tongue

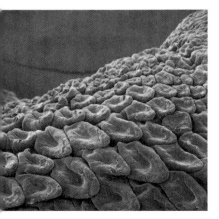

When seen magnified by a scanning electron microscope the papillae themselves look like a mass of small tongues on the tongue

the medial hypothalamus, which is involved in sexual activity, and the ventro-medial nucleus, which is involved with feeding control.

The Flehmen response of the cat family to the urine marks of other cats may be linked to sex hormones in the urine. The Tiger's Flehmen response is dramatic, as it appears to look disgusted, although clearly it isn't! The Domestic Cat's Flehmen gaping action, despite its importance to the cat, usually goes unnoticed by owners. Both toms and queens make the response.

A Tongue of Taste

Whenever a cat licks your hand, you are in direct contact with one of the cat's most useful tools – its tongue – and it is a rough contact! The Feloidea have the most developed bumpy tongue of all mammals. The upper

surface of the tongue is covered with a series of backward-pointing spines, or papillae, which give it many uses, from combing the coat and rasping its food to taking up more water to drink via the surface tension effects of the spines. The cat's tongue fulfils functions for which we use our hands. Consequently, the cat accepts stroking as the equivalent of being licked.

The cat does not have taste buds on the central spiny papillae, but on the more rounded papillae at the front, back and sides of the tongue. The tastes interpreted by most mammals are sweet, sour, salty and bitter, but cats alone show no particular response to sugars. The ultimate carnivore, the cat enjoys a meat diet, rich in protein and fats, but virtually devoid of sweet tastes. Kittens do receive milk sugar, lactose, when suckling from their mother, but lactose levels that cause no problems to newborns, can cause severe diarrhoea in weaning kittens. Milk is a major cause of diarrhoea in many cats.

A Touching Ability

The skin's specialised nerve endings pass on messages along the nerves on receiving contact pressure. These vary over the cat from twenty-five in a square centimetre of skin on parts of the head and feet, down to seven in a square centimetre of the back, tail or ears. The cat's nose, tongue and paw pads are most sensitive.

The Cat's Whiskers

We realise how special they are when we say something terrific is 'the cat's whiskers'. The cat's normal hairs, particularly the coat's longer guard hairs, are connected to touch detectors, so it is small wonder that after you stroke a cat it combs it back into place with its tongue! The whiskers are enlarged hairs especially endowed with sensory receptors.

When a cat yawns the muzzle whiskers sweep forwards in an array

When snow first falls some cats go wild with enjoyment as they play with the strange stuff, regardless of it being cold and wet

Those around the mouth feed information about the width of openings, which is as vital for going through a fence hole in the urban jungle as in dense woodlands when prowling around after dark. The muzzle whiskers are part of a circular facial array of whiskers, with superciliary tufts above the eyes, and genal tufts on either side of the face behind the main whiskers. In the dark we put out our hands to feel our way, while cats, because they are leading with their heads, use their whiskers.

Unlike other carnivores, the cats have no tufts under the chin. The gain of extra power in the bite of a cat by the foreshortening of the face length has left no room for such a set of whiskers. The most unnoticed whiskers are on the underside of the forleg, but these are vital for the arboreal cat, to judge landings. They also help the stalking hunter and aid prey catching with its paws.

The cat's expressions are enhanced by its muzzle whiskers. During yawning they sweep forward, while when the cat is being anxiously defensive they are flat back against its face.

TEMPERATURE DETECTION AND CONTROL

Cats love warmth. The cat curled by the fire is taken as a sign of contentment, yet, ironically the curled up cat is a cool cat trying to control heat loss. The true, warm, contented cat by the fire lies out long to lose heat! My cats often sit nearer to the kitchen range than I find comfortable, for while I can only tolerate a temperature up to 115°F (44°C), cats are happy to stay up to 125°F (52°C). The cat's nose and upper lip act as its key external thermometer.

Acclimatisation experiments have shown that the cat can habituate over a wide range of temperatures. It has shown this ability by licing feral, generation after generation, from the Equator to the sub-Antarctic circle. Cats have a thicker winter coat, and the transition to a thinner summer coat is dramatic in breeds like the Angora and Van.

The colour points of the Siamese coat is temperature dependent, at lower temperatures more pigment is made. As the points are cooler than the body they are where the colour develops.

Our cooling control is by the evaporation of sweat. Cats cannot have watery sweat all over their fur as it would increase the evaporation surface, so their watery sweat glands are restricted to the bare skin surface of their feet, around their nipples, and lips.

The dog, as an animal of pursuit, cools his body temperature by excessive panting. In contrast, cats, who primarily work by stealth, try to avoid becoming overhot in the first place, and strenuously avoid too much exercise for its own sake!

CAT CLAWS

An essential part of the felid make-up are the cat's claws, which, like a fistful of flick-knives, are protractable, not retractable as is so often stated. At rest they are sheathed and have to be actively protruded. Consequently, unless the cat is old or its claws are very long, any clawing of your furniture is intended! All cats can protract their claws, and share this ability with some other arborial carnivores. Cats sheath their claws to stop them being blunted when walking on the ground but keep these climbing spikes ready for use. This has given them the added advantage of available hunting and fighting weapons. In contrast the dog family has blunt claws as it walks on them all the time.

When cats catch prey, unlike dogs, they first engage with their claws. The small cats that pursue small prey initially make contact with paws and claws extended. Only when the prey is subdued do they bring their head forward to bite.

THE SUPREME HUNTER'S DESIGN
Branching Out – The Tail and Other Parts

The cat's design and build is that of a woodland hunter. It moves effortlessly amid tree branches, aided by the balancing movements of its tail.

The fastest cat, the cheetah, uses its tail as a gyroscope to turn at speed without falling over. Our cats use the same technique as they clamber about – if balance is threatened, the tail comes into play.

Contented cats, tummies exposed to the heat from the range. Lying out long is a clear sign of a warm cat

The cat's whiskers help it to judge close distances as it climbs, aided by its specialist's protractable claws

The Sense of Balance

In the inner ear, arising from the fluid-filled cochlea, are three semicircular canals. These canals are at a 90° angle to each other. When the cat makes any sudden movement, the fluid tends to stay put, and relative movement is detected by tiny hairs. This gives the cat its sense of balance.

We have a similar system, but unlike the tree-clambering cat, we do not have a self-righting reflex. As cats can fall from trees, this rotating mechanism is a definite survival advantage. The cat's ability to rotate rapidly when falling is incredible, and reveals a highly attuned gravitational detection, for it needs to know when to stop rotating. This information is partly provided by sight, but also by minute pieces of chalk in the semicircular canal system which land on the haircells from the fluid.

Time

The cat's sense of time is remarkably accurate: if you feed your cat at a set time, it will turn up virtually to the minute. However, if cats are put in continuous darkness, their internal biological clock begins to drift. Morning light seems important for resetting this biological clock.

ESP AND PSI

Many people want to believe that cats have extra-sensory perception. It is strange that we want the cat, with its finely honed senses, to have yet more, but as it can 'taste scent', why not? Many believe cats have super-navigational abilities. Birds migrate in part using magnetic detection. Bees navigate via the sun's position. So if the cat does have good navigational ability it is less likely to be extra-sensory perception, and more likely to be perception via an extra sense. But do they have it?

I am frequently told of cats finding their way back to their old homes when their owners have moved. When I have investigated, it seems cats have found their way back to the old home over unfamiliar ground, at least over relatively short distances. There is no need to evoke magic, for if the cat's attachment to its home range is strong, then it may feel the need to return. As other animals can navigate via natural phenomena, so perhaps can the cat.

What is harder to believe are the rare claims made that cats that have been left behind have been able to navigate to their owner's new home. That does seem impossible unless people send out 'vibes' like homing beacons! Yet one New York vet moved across the United States to California and some months later a cat just like his old one turned up. He was sceptical until, he claims, he found the cat had an identical bone injury to his old cat. The term 'psi-trailing' has been coined to cover such almost impossible journeys.

The arborial cat has a ready ease while clambering at height among small branches. Its sure balance is greatly helped by its tail

71

Sex – the barbed question

THE SEXUAL CAT

The cat has always been synonymous with sex. We still talk of a sexually demanding woman as like 'a cat on heat'. The cat's dramatic sex life made it the perfect candidate to become the fertility goddess Bastet in Ancient Egypt. According to Aristotle in 350BC, 'the female cat is particularly lecherous, and wheedles the male on to sexual commerce'. Aelian in AD120 noted that 'the male is extremely lustful, but the female cat is devoted to her kittens, and tries to avoid sexual intercourse with the male because the semen which he ejaculates is exceedingly hot like fire, and burns the female organ'.

The witchcraft era held to the belief of 'seed of fire' but interpreted the cat's apparent sexual excesses as wantonness and wickedness. The reality is no less dramatic. Female cats go through several cycles per year of sexual responsiveness. Breeders' cats that are housed or cats that live in warm climates will come into season earlier in the year than in the more northern climates. The timing seems mainly to be controlled by light levels. Cold weather and low light physiologically dampen the ardour of street cats. This is no accident, for post-weaning survival of kittens can depend on the mother's ability to catch sufficient prey to feed the young.

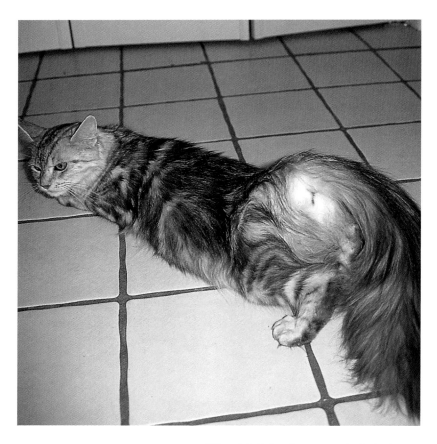

(Above) A Maine Coon queen in the lordosis position soliciting the attention of a male, demonstrating her readiness to mate

(Top right) A queen rolling 'on heat'

As she comes into season, the female's behaviour changes and she begins to call and advertise her state. Her sexual scent and sounds encourage interest from toms, but initially she will not accept their approach. Her oestrus state lasts for up to four or five days, during which she rubs her head and chin on objects, rolls about and crouches low with her rear raised and rhythmically paws the ground.

When the queen's hindquarters are raised – the position termed lordosis – the male makes chirp-like quiet calls to her, and can increase her responsiveness by licking her vulva.

The male often needs the reassurance that he has a right to be present, and to mate successfully some toms need time to establish 'residence', in part by scent marking. A young, inexperienced male can be intimidated by an aggressive queen, such that his first successful mating can be delayed by months.

While the queen rubs her head on the ground, the male approaches indirectly, moving around her. He moves and normally grasps the loose skin at her neck. He then mounts adopting a hugging grip with his forepaws around her body. Copulation at a breeder's may take only some ten seconds, but in free-living cats can take $1^{1}/_{2}$–2 minutes.

Edward Topsell expressed the event in the seventeenth century in an unusually tolerant way for the time:

The manner of their copulation is this, the female lyeth downe and the male standeth, and their females are above

While mounting, the male secures a firm hold of the queen by grasping her by the scruff of her neck

measure desirous of procreation, for which they provoke the male, and if he yeeld not to their lust they beate him and claw him, but it is only for love and not for lust.

The one-to-one relationship between male and female that occurs in the confines of a breeder's cattery is far from normal; when cats are free-ranging the scents and calls of the queens cause a number of toms to gather. The friction caused as toms have to pass through each other's territories has long been noted. Bartholomeus Anglicus wrote in the mid-thirteenth century:

In time of love is hard fighting for wives, and one scratcheth and rendeth the other grievously with biting and with claws. And he maketh a ruthful noise and ghastful, when one proffereth to fight with another.

73

A scanning electron micrograph of the male cat's penis showing the spines that trigger ovulation in the queen

This is based on shrewd observations, for the caterwauling that many attribute to mating, occurs earlier and is due to the threats that prelude fights. The nocturnal 'cat-song' in suburban areas is just as likely to be straightforward territorial disputes.

However, there is a scream that the queen emits which is at the heart of sex in cats. Unlike women and most other female mammals who release an egg as part of their regular sex hormone cycle, cats share with a few other carnivores the need for stimulation to induce the egg's release. Male cats have a small bone in their penis called a baculum, but the particularly unusual feature of the organ is that its flanks are covered with backward pointing spines. These also occur in other carnivores that are also induced ovulators. I found that on the cat there are between 130 and 190 spines, and the more sexually mature the male the larger the spines. Neutering reduces the androgen levels and so the castrated male's spines regress. The spines are the anatomical feature that caused our ancestors to talk of 'seed of fire', for they cause the female to react violently and scream at the point of withdrawal. The barbs ensure that mating with the first tom triggers the release of the egg, which then travels down the fallopian tubes until it reaches the site where it can be fertilised, a journey which takes many hours and during which the female cat mates with numerous toms.

In many mammals mistiming of mating and ovulation is a major cause of infertility. In the cat the male's curious anatomy linked to the female's need for induced ovulation avoids this problem.

The cat's wild ancestors were spread out by their elaborate territorial system, and this way of mating bought time to allow toms from different areas to congregate and so avoided the double problem of missing ovulation and inbreeding within a group. This is the system our cats have inherited. Ovulation occurs around 24 hours after the first mating, and fertilisation follows at between 24 to 36 hours post-copulation.

Few studies have been made that look specifically at mating within large groups of free-living cats, but a recent one by Eugenia Natoli and Emanuele Dé Vito on the groups of feral cats living among ruins in the centre of Rome has cast light on the secret sex life of the cat by following the queens that came on heat. During the course of a queen's oestrus, she would be courted by eight to twenty males, and although at any particular mating session there would be up to fifteen males around her, five was most common. Most females copulated with more than one male during their oestrus period, some with up to seven males.

An important feature of cat mating that emerged from this study is that significantly there were 2.5 times more mounts taking place without intromission than those with full mating. The average was 29 without intromission and 11 with intromission for a queen in 24 hours. For the average queen that amounts to 39 sexual acts in

24 hours, or 1 every 38 minutes. Small wonder that the ancient Egyptians made the cat the female goddess of fertility!

I found in my study of East London housecats that some toms will make appeasing overtures to the early pro-oestrus queen to try to move closer to her. She will allow his approach to a certain distance, but then stop him. This different strategy in contrast to charging directly in has also been found in the Rome study. Although the Italian researchers found it did not seem to matter which strategy was used with regard to the outcome, from my London observations it seems that less assertive males may successfully pursue their end via this route and minimise the risk of fighting with dominant toms.

After mating, the female normally rolls onto the ground, and both male and female will separately wash.

When cats are in either the controlled conditions of a breeder, or there happens to be only one male around with free-ranging cats, the male initially makes all the running, while the female is not that keen. As time goes on, he becomes more tired, while her enthusiasm grows. When more males congregate, which is the more usual situation, a more even balance is achieved.

The apparent nymphomania of the female cat is concomitant to the species' territorial patterns, which themselves emerged from the way of life imposed on this carnivore by its forest ancestry. Its behaviour, design and way of life mesh as a survival package. The multiple matings in addition to ensuring

A traditional build Siamese queen in Thailand with new-born kittens. As they grow older they will develop the colour on their points

fertility give further advantages as they can produce different sires to a single litter, increasing genetic diversity in a group. Cats that live with and around us in urban settings live at fantastically higher densities than they would if we were not on the scene. Consequently, more males can be present at a mating than could have occurred historically, so genetic diversity is being accelerated. This has been offset in recent years by the increase in neutering in the town housecat and feral populations.

THE BIRTH

If you have had your cat mated and she conceives, she will miss her next heat and then her nipples will redden and become more pronounced. At three to five weeks after mating a vet should be able to check for pregnancy by very gently feeling her abdomen.

As the embryos turn into developing kittens, they will become a nutritional

75

A mother cat purring and rhythmically pawing as her two-day-old kittens suckle

drain on the mother and she will gradually feed more from around the fifth week of pregnancy. From six weeks onwards her larger size will be noticeable to the eye. As her size increases, she will spend more time resting, and will often lie on her side, as she will do later when suckling, for this spreads the weight of the developing kittens.

Overall, her pregnancy will last around nine weeks. As her term approaches, she will begin to look for a kittening spot, and probably will accept the offer of a cardboard box cut down on one side and with newspaper bedding. This should be put in a warm, quiet, dark corner. However, she may prefer her own choice.

Virtually all cats manage to give birth with no problems and without the need for human intervention. Certainly advise your vet of your cat's condition, but a

vet will only need to be involved if the cat becomes distressed. In a feral group or multicat household, other females may assist. Queens often purr throughout the process of giving birth.

The queen will lie down in a semi-prone position, with her upper rear leg partly raised, and a vaginal discharge will occur. This can go on for some time, but once contractions start and she begins to strain, she should give birth to a kitten in its sac after fifteen to thirty minutes. The mother's rough tongue will lick away the sac and wash the kitten. Her licking will also stimulate the kitten to take its first breath of air. She will cut through the umbilical cord herself, eat the afterbirth and clean up around the kittens. The time between the births can vary considerably, and although the average number of kittens is four, this can also vary.

It was suggested earlier this century

that as housecats can have up to three litters per year, while the Forest Wild Cat *Felis silvestris* typically has only one, that this constituted a clear difference between the Domestic Cat and its main ancestor. However, I found that by analysing the trapping data of feral cats and kittens of Stichting De Zwerfkat in Amsterdam, one litter per year seems to be normal for feral cats, with the peak being in April and May. It seems that it is not domestication but domestic living that allows the house cat to have late litters, and that living wild the Domestic Cat is more like the Wild Cat in normally having only one litter a year.

TIMETABLE OF KITTEN DEVELOPMENT

It is impossible not to fall in love with a little kitten. When people go to a cattery and see some kittens, they often cannot resist taking one home. The reason for this is that as well as looking small, fluffy and cute, they have a flatter face with proportionately bigger eyes than their parents. This appeals to us, for they are the features of our own babies.

The First Two Weeks

There is a definite timetable for kitten development. When kittens are born, it is the beginning of a big adventure in which at first they are totally dependent on their mother's support. In the first one to two days, the kittens receive the vital colostrum from their mother which gives them their initial immunity, after which they thrive on their mother's normal rich milk. She feeds them completely for the first few weeks.

The new kittens have to find food, despite the fact that their eyes are shut until day five to ten, and that they have minimal hearing until day five. They navigate their way to the teat by their sense of smell, and also by homing in on warmth. The mother helps by lying on her side, exposing her teats, and nuzzling the kittens in the right direction. Their sense of touch has been with them from before birth in the womb.

While puppies have little loyalty to any teat, kittens do develop a preference. This is an advantage for the mother as even young kittens have sharp claws which could damage her teats in a squabble. Even at a very young age, the paws open and the claws protrude as the kitten 'paddles' while suckling.

When kittens begin to feed, they engage in long, eight-hour suckling sessions. From very early on, they will squeak to the mother to let her know that they want food or have strayed

A litter of ten-day-old kittens at the 'blob' stage huddled together for warmth. One is making a defensive hiss outwards

out of the nest huddle, they work their way back in, pushing their heads into its midst. Their small legs are not strong or long enough to lift their tummies off the ground, so they drag themselves in a wobbly way back to the warmth and security of the others, or find a teat.

I find that during the first two weeks, before their eyes bring them focused information, the kittens, with their eyes shut or open, will open their mouths wide and give a defensive hiss when a strange scent of an intruder is directly in front of them.

Two Weeks Old

When the kittens' eyes first open, they are very cloudy and bring in little information. During their third week, although their walking movements are still shaky and they stay as a huddle, for the first time they will sit up on their haunches properly and *look* at things. If you are around, some of the kittens will look *at* you, and they certainly recognise their mother. The automatic defensive hiss disappears. Some eye fluid cloudiness will remain for another few weeks. Their ears are still small in proportion to the round face at these early kitten stages.

Three Weeks Old

When kittens reach three weeks old they begin to explore the world. They discover how to walk, placing their paws down in a fairly shaky way. As they begin to move around, the littermates enter properly into the activity of play.

Top: At two-and-a-half weeks the kitten's ears are still small, and its shaky legs cannot lift its tummy clear of the ground

Above: The kitten at four weeks old, with longer ears and legs strong enough to stand at full height

away from her into the cold.

During their first two weeks kittens seem like cute, little, unsteady blobs, for their heads are quite round and their ears quite small. They have a strong desire to stay as a heap huddled together in the nest when the mother steps out. Their noses are sensitive thermometers and if they are

Five Weeks Old

At four-and-a-half to five weeks kittens are more robust. If you sit with them, they are able to clamber up your clothes, albeit in a slow and lumbering way. They begin to mountaineer, but may need some help! Their vision is now good. The cloudiness in their eyes has now disappeared completely, and these have become sparkling clear. They had some hearing on day one, but now it is very good. Hearing will continue to improve until they are twelve weeks old. By five weeks old, they are gaining that much more information on their world, and they make more sense of life.

At this age, the excreta of feral kittens would normally begin to change from that of just a diet of milk, to that which includes the occasional vole. In cats that live with people the change is to cat-food.

While the mother cat has been content to lick away their mess from the rear end, its new constituent changes her mind. For survival, it has mattered that up to this stage the mother prevents the appearance of detectable mess around her nest while the young are too small to be very mobile. As the mother's behaviour changes the kitten no longer needs her stimulation to eliminate.

The play of five-week-old kittens is much more firm in movement. They arch and thump down more determinedly on littermates. They jump up and down and bat their fellow kittens quite heavily.

Seven Weeks Old

At seven weeks kittens are usually weaned and are eating solids; suckling ceases. They are often taken away to their new homes at this stage. The kittens are now more self-possessed and they are much more co-ordinated. In their play, they turn to pursuit games as if they are after prey, and they now play more with objects. Their improved sensory input lets them have better

Top: A Maine Coon kitten, bright eyes and large ears, just weaned and dashing about

Above: Two black-and-white, five-week-old kittens still suckling. To help stimulate the milk flow they open and close their paws, paddling at the teat

79

Eight to Fourteen Weeks Old

Once a kitten reaches ten to eleven weeks old its motor co-ordination is fully developed, so potentially it has the flexibility and movement control of an adult.

From eight weeks onwards kittens are beginning to react like adult cats towards a threatening situation, so play begins to change again. At around nine weeks old, play between kittens becomes more earnestly aggressive, and squabbles can become spats. These are still rehearsals for the adult world, so there are lulls in the rushing around which let the 'protagonists' get their breath back!

Until kittens are around twelve to fourteen weeks old the level of playful interaction between kittens remains high, and a lot of it is as much about object pursuit, running around objects and generally investigating, as it is about scrapping. By fourteen weeks old, the aggression has seriously weakened the youthful bonds and the litter dissolves as adulthood opens before the now independent juvenile cats.

Sensitive Periods: Learning For Adulthood

Sensitive periods used to be called critical periods, implying that there was a once-and-for-all learning period. Now, however, biologists recognise that cats can also learn later in life, if not so well. Nonetheless it is better if the kitten can learn at the right stage, otherwise it can take much longer.

One such window of learning is

Five weeks old and beginning to clamber – but not too well!

body control, so their play noticeably changes. The mother's behaviour also changes as weaning progresses. She no longer lies back exposing all her teats, but begins to roll over slightly to cover up her nipples if the kittens are still demanding. Their reflex to right themselves if they fall off something has now developed, just in time to be useful as life has become more of a scramble!

around four weeks, when the mother begins to bring home prey for the kittens. Up until this stage, all they have encountered is their mother's milk, so it must seem strange to the kittens suddenly to see a dead vole. But during this sensitive period kittens are very trusting – whatever mother gives them, they are going to believe is good for them. The kittens need to have this acceptance and trust of whatever mother brings in, otherwise when she brings back live prey they might become frightened.

The mother starts by bringing home dead prey, then after a while she graduates them to subdued, injured prey, and eventually to lively prey by the time they are weaned. She trains them by acting as a competing kitten. Their interest is stimulated and they learn about prey by handling it. Her apparent competition then increases.

Handling Kittens

Sigmund Freud would be pleased that the cat world is beginning to catch up with his pioneering work of evaluating the adult from what happened in childhood. Even today, in most catteries, whether those of breeders or of cat rescue, there is little time set aside for handling young kittens. Much of the effort put in by breeders to get the appearance of the cat right can be negated by paying no attention to kittenhood behavioural development. What usually happens is that people service the enclosure of mother and kittens by providing food, changing the litter and keeping it clean. Particularly

during the 'blob' stage of up to three weeks, handling the kittens can seem like invading their sanctuary.

In reality, that is exactly what is happening, for among cats living feral the mother will make strenuous efforts to avoid having her kittens put at risk

Below: A six-week-old kitten in classic position demonstrating fear/aggression: fluffed-up full tail and arched back, side on to the other kittens

Bottom: The rough and tumble of boisterous play among seven-week-old littermates

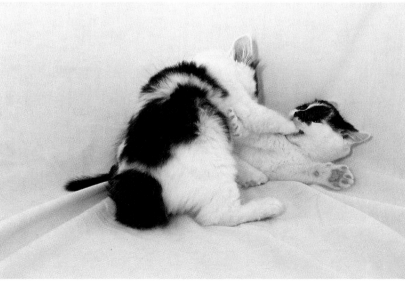

81

Seven-week-old littermates competing over a mouse

by anything, and she will move them whenever she feels it is necessary. However, despite this strong protective urge, it has long been known that mother cats will bring up other animals in their litter alongside their kittens. Harrison Weir noted in 1889: 'They [cats] will rear other animals such as rats, rabbits, squirrels, puppies, hedgehogs; and, when motherly inclined, will take to almost anything, even to a young pigeon.'

If prey species are brought into the nest area by the mother in the normal sequence of dead to an increasingly live state, from when the kittens are five weeks old in anticipation of weaning at seven weeks onwards, the kittens will recognise them as prey. However, before that time in the biological timetable, intruders should not be in the nest, so if they have been planted by devious means then the kittens accept them as clearly the mother has allowed them into their 'safe haven'. The kittens behave towards such co-habiting animals as fellow littermates, and, as

was shown by researchers over thirty years ago, they will continue to behave differently to that species for the rest of their lives.

Our intruding into the kittens' nest and handling them should be seen in that perspective. When we handle kittens and they habituate to people, it is not due to some magic influence of people as such, but we can take advantage of the kitten's development that will help to put them more at ease with people in later life. Eileen Karsh in Philadelphia in the late 1980s began to investigate the differences that earlier handling would make. Compared to kittens with no handling, she found that post-weaning handling made little difference to their friendliness when they became adults. In contrast, earlier handling from between two to three weeks until weaning at seven weeks, turned kittens into much more sociable cats with humans later in life. This would seem to be the cat's sensitive window for socialising with people. Although kittens should not be removed from their mother until weaned, they should be handled by people earlier than is customary. It is better that such gentle handling should be in the nest area to avoid causing separation anxiety and later behavioural problems.

If you go to a cattery and bring a kitten home once it is weaned, you will naturally lavish love and attention on it. Unfortunately this will probably be the first time that your kitten will have experienced such handling, and it will be too late to condition the kitten fully to people. Well before you go to any cattery, ring up and enquire whether, as policy, someone spends some time each day gently handling the kittens in their nest area. It is certainly not a good idea to remove a kitten from its mother pre-weaning for that in itself will cause behavioural trauma and they can become fearful and aggressive with people and other cats. But time spent by people with pre-weaned kittens in both breed and welfare catteries is not wasted time.

Habituated kittens will fit even more comfortably into human households for the rest of their lives, having discovered that we are not monsters!

SEXING KITTENS

When you are looking for a kitten for your household, it is fairly important to *know* its sex, and not just to be told 'what it is'. It is not uncommon for Tom to change to Tabitha on discovering after a while that the advice was a bit slack!

The basic rule is that when you look at the rear end of a kitten the apertures of the vulva and anus are close together in the female. In the male, the apertures, this time of the penis and the anus, are further apart to allow for the position of the testes.

If you have any doubts as to what 'close' and 'further apart' mean, look at a few of the other kittens in the litter and you will soon recognise the relative distances.

Six-week-old Abyssinian kittens, still nestling together and at the most appealing of kitten ages

The territorial cat

Secure in its territory, a house cat sleeps in one of its preferred sunning spots – and by its presence and way of 'nesting' declares its ownership

Uniquely among domestic animals, the cat defines its own use of land, while with cattle, horses, dogs or gerbils we tether, fence or cage them and impose their area of use. However, we have no need to contain cats, for they know where the edges and limits are on their own ranges. The cat determines the edge of its range as a consequence of a number of factors.

In everyday English we often use the word 'territory' to cover anything connecting an animal with its land, yet technically to the behaviourist, a cat's territory is that area of land it will defend against other cats. The area that it uses on a day-to-day basis is its home range, and although it is often quite similar, invariably it is larger than the territory. In reality, neither are fixed entities, and the area utilised varies at different times of year and also depends on other changeable factors, such as a new cat next door!

It is not just cats who play this game, for we recognise our own territorial boundaries by the hedge or fence between our garden and the neighbours. Yet we range further as we travel about our daily lives. The cat does not consider a fence by itself a territorial limit. A cat living on the other side of the fence will change its significance!

In rural areas, a cat's range can be large as there are few competing cats about. In a suburban landscape cats live more closely packed together. However, in essence there is little difference in the pattern of land use, just its size.

To the ecologist, the key factor about the home range is that its size has to be sufficiently large to provide the food needs of the animal. So the richness of the pickings in the landscape dictates how far a wild cat or a feral cat has to range to survive. Consequently, the carrying capacity of rural hillsides for feral and farm cats is much lower than that of urban and suburban areas where densities of house cats are high due to the superabundance of food. As a result, a well-fed suburban cat does not need a large range, and the presence of neighbouring cats ensures it is confined! In that sense, territorial considerations set the limits on the size of a cat's home range beyond that of the minimum area for food requirements. The boundary division between two female cats can be recognised by the way they sit on their own side, and stare at each other.

Suburban queens can hold parallel ranges in a series of adjacent gardens. Toms are different. An intact tom may range over an area up to ten times the size of the range of a local queen.

A Spanish cat wakes up after a siesta and stretches, before setting off on the rounds of its territory. It shares the core area of its range with other cats around the house

Between sleeping, life is an earnest business of watching for territorial intruders

1

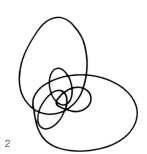

2

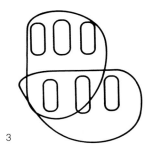

3

The patterns of home ranges of cats: the larger loops represent male ranges and the smaller loops those of females.

1 The normal wild cat range pattern, and also the feral cat range pattern where food is dispersed and scarce
2 The clumped ranges of feral cats around a locally abundant food source
3 The ranges of house cats

CAT RANGES

The cat's semi-social life works on the balance of food and range, and one dictates the other. If a cat has a ready source of food, it does not need to travel so far for food, and consequently does not need a large range. In the Highlands of Scotland, with little food to scavenge and most having to be hunted, a feral cat needs a huge range. At the other extreme, the density of cats is high at the centre of towns, even for feral cats who can scavenge from surplus food in bins and plastic sacks as well as receiving handouts. The highest density of cats is achieved by house cats for we virtually pour food down their throats so they do not have the need for much space.

The densest area in the world for cats is suburbia, with small back gardens. The remarkable feature is that from range studies of Tigers and European Forest Cats as well as Domestic Cats living feral, the overall land-use pattern for the cat family is consistent. Yet why is the male's range up to ten times larger than that of the female? It does not need to eat ten times more food than the female, for although it is larger, it is not *that* much larger! The reason is social. The female is the basic unit and her range is the minimum size that can support a cat. We often assume the male cat is idle when it comes to rearing young. The female Tiger and tabby first suckle, then bring back food for their kittens without any apparent involvement from the male. However, the male does expend a massive amount of energy patrolling and

A territorial dispute between neighbouring neutered house cats at their boundary fence. The female aggressor has fluffed up her tail and back, while the male is standing sideways on with arched back to appear larger. Like most disputes, it did not become a fight, and all moves were in slow motion. Its significance is that it shows to each cat the current willingness to retest and redefend limits that are fluid

protecting the huge range within which a number of females may rear their young.

If food is spread out and thin on the ground, female feral cats, like wildcats, do not overlap their ranges. Yet if there is a rich local source of food they will readily clump their ranges around it. The toms will also clump their ranges but wander further, and still give a protective buffer zone to their group's queens from other cats or cat groups. This is what happens around a group of dustbins or where cats are fed in an urban square by a feeder.

Our relationship with our own house cats is also governed by this ratio and pattern of land use. If you

live in a house with a queen she will identify with your use of your garden. Other queens identify with their owners as individual groups in a series of parallel ranges along a stretch of gardens. If you own a tom, it will hold a larger range than the local queens. It will recognise their areas, and then range relative to them up to ten times the area if he is intact and up to about seven times if neutered.

The tom's range will naturally overlap a number of queen's ranges. But whereas historically, or in a feral state, a tom could expect that the queens he overlaps territorially would relate to him, in the suburban landscape of house cats, each of the queens will identify with their owners as group members and perceive him as an outsider – modern suburban life can be hard on the male of the species!

Paths and Snoozing Spots

Whatever the size of your garden and your house, there are places your cat just does not visit, for it has particular paths. It will also have preferred snoozing spots. Cats take great care with their temperature control (See pp68–9) and move between favoured warmer and cooler places within their range.

When basking in the sun, cats manufacture vitamin D from the action of the sun on 7-dehydrocholesterol on their coats, which they ingest when they groom. The cat confined indoors does not usually have this advantage, but will sit in the sun as it filters through the window glass. However, it still alternates between warm spots, such as by a radiator or in the airing-cupboard, and cool spots, such as behind the sofa. For cats, particularly on cooler days, one of the advantages of sitting on people's laps is that they can absorb the warmth!

As winter approaches, cats progressively spend more time in the warm indoors, and as a result put less effort into maintaining the size of their exterior territory.

Snoozing indoors or outside, the cat is ever watchful of territorial incursions

YOUR PLACE IN YOUR CAT'S GROUP

In many ways I behave towards my cats as if they are human – and I'm not alone in that! Similarly, to my cats I am some sort of cat. We are all a bit confused on the overlap between people and pets.

When you look at how your cat forms its range, it becomes clear that it relates to you as if you are a cat. Our cats do not look to us to build defensive barricades, so while they won't regard a fence as a limit just because it's there, they do look to see where and how we use space. If you never go into your garden, your cat may well have no particular attachment to it. If instead you love to sit in the garden, take afternoon tea and bask in the sun, you will have a companion alongside. Nothing will bring a cat to sit beside you so quickly as doing a spot of weeding – the curious cat will come to see what you are doing. You are sharing an identity of land usage. While a fence may not demonstrate ownership to the cat, it recognises your presence as declaring ownership and gains confidence of its right to be there. because you are there.

Scent Marking

Cats, being small animals, do not have the advantage that we take for granted of being able to see over things into the distance. Our gardens, and both rural and urban environments, can be as much a jungle to them as the landscapes of their ancestors. In consequence, the ability to leave scent marks around their range helps them to read the landscape to see if it is unassailably theirs, or whether other cats are about. Scent marks are one of the ways that other cats know where your cat has territorial presence, even when it is not around. As a cat moves around its range, it has some indication of time from scent marking: not only does scent age, but other cats may have overlaid or intruded on the scent since the cat was last there.

A cat does not just leave scent trails around its range but on other group members as well, and that includes us. By rubbing against us, and when we stroke them, reciprocal scent messages are left which are non-threatening and behaviourally supportive within a cat group. We may be unable to interpret such scents, but the cat's life involves checking scent directly, and by taste when washing, all the time. The Jacobson's organ over the roof of their mouth (see p65) is able to aid such investigation. You may well have noticed that your cat will investigate your shoes and trousers very closely with its nose almost glued to them after you have returned from somewhere that had other cats. Your cat may well suddenly stop, and go into the held gape of a Flehmen response as it allows the scents to be brought into the organ and interpreted. This happens in both sexes of the cat. Our return into the house with a cardboard box of shopping similarly elicits a full inquiry! If cats are interpreting such smells as originating within their range, or just out of it, we must cause them some consternation.

There are scent glands on a cat's head and on the top of its tail near the body, so when a cat rubs against us it is leaving a scent. However, it also has huge sebaceous scent glands along its lips and chin. (If a cat develops acne

on its chin or 'stud tail' the glands are blocked or overproducing wax.) When cats walk around their range and leave information about their presence, they use these glands. The action by which they frequently and carefully rub against a small twig or dried stem of grass sticking out into their path, is called 'chinning'. This movement as the cat brushes against objects can be almost

Top: A neutered queen 'chinning' on the ground in response to finding an 'interesting' scent there

Above: A neutered female house cat checking out a 'scent-stick' that another cat has marked by brushing against it

imperceptible, but it is enough to leave a residual scent – a message. If you have not noticed these extremely significant signposts, squat down to cat height and they will become obvious – obvious enough that other cats recognise them and go straight to them to check them out.

If you watch your cat as it goes out of the door or through the cat flap into the larger world, it does not usually charge out but moves carefully checking as it goes. As a cat passes through a gap in a fence, or through its cat flap, it will leave a deposit of grease from the coat which other cats will check out by sniffing.

The linkage of 'group identity' between ourselves and our cats can sometimes be 'one removed' in transferred behaviour. When a shy cat wants to rub against us but is too nervous to come too close, it will carry out the same action by rubbing against an adjacent object.

Broadcasting a Message by Spray

The pungent scent-spray of a tom broadcasts its presence widely. As the position and shape of particular twigs elicit the chinning response, so certain shapes and positions, such as a low wall, elicit spraying from an intact tom. As intact males are often intent on asserting their territories they will especially mark by spraying key strategic places that may be disputed. Consequently you may find your cat flap sprayed by a local tom as it detects the scent of your cat from the aromatic coat oils accumulated on the flap. If you have a timid cat, it will find such a situation quite stressful, particularly if it has previously been attacked by an intruder. Hub caps on cars are frequently used and checked by cats because of their shape and size and because they are at a comfortable height for the tom to spray against.

Although an intact tom's spray is strong enough to be read at a distance, at times other cats will go up to it and give a Flehmen response. While spraying by an intact tom is made by juddering against a vertical object such as a wall, neuters and queens are more likely to spray against a horizontal object. However, when asserting a right to be in a territorially contested spot, a neuter may well spray vertically like an intact tom, but its spray will lack the smell.

Scents are particularly significant to other cats, for not only do they give the right of presence, but they also convey information on sexual condition. As the spray can also be a bit strong for us it is one of the main reasons that owners have toms neutered.

Opposite top: Territorial marking of branches on a tree in the cat's area by rubbing with the side of its mouth, leaving a scent from skin glands and saliva

Opposite below: A Maine Coon working farm cat in Maine checking a rubbing point to see if another cat has marked it since it last marked it itself

Below: Spraying is universal throughout the cat world. Here a cheetah sprays in exactly the same way as would a domestic tom cat

Territorial Clawing

When a cat rears up against a tree with outstretched paws and drags its claws through the bark, a number of events are going on, one of which is scent marking. Certainly the cat is sharpening its claws, for by dragging them through the bark, moon-shaped shards of keratin splinter on each side, bringing the claw to a point. Cats are fussy about the type of tree they will use, preferring something like a young fruit tree or established pine tree where their claws can sink in and the bark is soft enough for them to be pulled through. Claws are blunted by cats walking on our concrete-clad landscape, so they need to re-sharpen them.

Although cats enjoy the action of stretching, and clawing on a vertical object, they will also do the same on fallen timber. The action is that same characteristic move after a cat wakes from a snooze, takes a couple of steps forward and while keeping its hindquarters raised, it will stretch out its front paws and make a long curve of its body. The only difference is that the clawing is added.

However, this action is multipurpose, which is appropriate for an animal that does not like to waste effort! Cats have very few sweat glands on their body, for being a small furry animal they would lose too much heat on its evaporation. They do have watery ecrine sweat glands on their feet which keep the skin supple so they retain touch sensitivity in their pads. As nocturnal hunters, they need information on the nature of the surface they are walking on. They are as cautious of walking on thick mud or gravel as they are of walking on thorns. When cats claw a tree, they leave faint scent messages from their sweaty feet, which would be ignored by other cats, but for the visual clawmarks that draw their attention, and they sniff these.

Although clawing is commonly carried out against a vertical tree, it is often done on fallen timber, or as here on a horizontal fence rail

94

Dropping the Scent

Many animals use their droppings for territorial scent marking and the cat is no exception. It is unique among domestic animals in burying its faeces, but not always, and in that lies a flexible scent marking role. If you watch your cat when it is covering up its droppings, you will see that it carefully checks the scent to ensure they are smothered. Strangely, covering does not destroy scent, but extends it by preventing it drying out quickly; at the same time, however, it does make it more discreet.

Dominant toms, though, often leave droppings exposed in areas of dispute in the manner of a fox to assert their authority. It need not be just a dominant tom but one sufficiently confident to claim a presence. As an example, my neuter tom Leroy developed an infected abscess following an attack by a dominant tom. On his return from the vet he was kept in for a week to recover, but during the day I took him for a walk on a lead through a field which his rival had claimed. For the first few days Leroy deposited his dung very visibly on top of flat grass knolls until he considered these were sufficient to claim a right and he subsequently returned to burying his droppings. During this same period, he sprayed vertically against bushes which he normally would not do and he chinned a lot. He also growled at his rival on sight and my presence ensured that he maintained his confidence.

However, when after a gap of a week I took him along the same route, for the first few days again he left his droppings prominently sited and made no attempt to cover them, in contrast to his normal behaviour within his own range.

Cats usually prefer to leave their droppings semi-peripherally to their confident range, but they are restricted by available sites. Lawns and concrete are not enticing, and gardens with weeds and rain-packed soil present few opportunities, but the next-door neighbour's fine tilth is ideal! His vegetable plot is excellent for both ease of digging and position. If you loathe cats and find your garden is a communal loo, your best defence is to have a cat of your own and your garden will be dug over less!

When cats dig they take their weight on their back legs, and alternating their front paws dig towards themselves, digging deeper as they go. In an average garden soil some twenty-two paw scoops are usual. When the cat is using its dug latrine it adopts a characteristic position with its tail up, at up to 50° from the horizontal. If, as a behaviourist, you have to be able to tell whether a cat is defacating or urinating, you do not have to be too indelicate, for the heaving sides are the clue to the former action! The cat will then sniff and cover, and repeat until it is happy.

Cats also have anal scent sacs on either side of their anus that are lined by secretory cells whose production is linked to hormone levels. These may add scent information to faeces. They are also the secret scent communicator, for when we stroke a cat or cats greet each other, their tails whip up, exposing the urinogenital area and the scents.

A queen in characteristic pose having dug a latrine in a neighbour's gravel path!

The hunter supreme

A mole lying on its back defending itself from the cat that has taken it back into its home. The cat's whiskers have been brought down towards the mole, and will give the cat instant information on contact with the prey

HOW THE CAT HUNTS

The cat employs different hunting techniques for different types of prey. For a bird in the garden it tries a variant of the dash-and-freeze party game. It keeps its body slung low so that its shoulder blades stand high like those of a cheetah. The bird has a range of advantages. It has a huge field of view, reaching a full circle in some species. It also has that most frustrating facility, to the cat, of flight. If that was not bad enough, most give shrieking alarm calls that put other birds on their guard.

Cats are most successful at catching our garden songbirds in the early summer when there is an abundance of available, ill-equipped fledglings. The parents are at their busiest and may take risks.

Young rabbits, in the first month after they come above ground at two months old, are a favoured prey. These are stalked and the inexperienced youngsters are easily surprised.

For small mammals, but particularly Short-tailed Field Voles that live on the surface of the ground under the cover of grass stems, the attention of cats is suddenly caught by their sound. Having homed in on the voles' position with its ear cones, and sitting stock still, the cat then rears up and pounces down with its front paws together in exactly the same manner as a fox. If successful it gathers up its prey straight away. If not, it thumps the place in the grass a few more times using smaller moves. Out of the grass the cat is unlikely to rear up, and attacks directly with its paws.

Small mammals such as mice and voles will seek cover for protection, but if forced out into the open, or taken by the cat into a more open patch and released, then they will run along the side of walls, rocks or fallen trunks. The cat will attempt to bat the animal with its paw to stop its escape.

Why Play with Prey?

Even the keenest cat owner is uneasy about a cat playing with its prey. The cat is superbly armed to deliver a death blow, so why on earth doesn't it just get on with it? If owners were cat-sized they would soon see why! Cats go in paws first with their heads well back. Only when they are totally sure they

The arch-hunter Leroy in action in pursuit of a Yellow-necked Mouse

are not going to be bitten do they bring their mouth into position. But with short faces they share with us the inability to see what is going on around the mouth. The neck bite that is a feature of the cat kill requires a very intimate positioning at the back of the victim's neck. If the prey moves, or, even worse, bites back, the cat is in trouble. Consequently it matters to the cat that the prey is not only fully subdued but stunned so it can safely make the kill.

Even a shrew can be a daunting opponent, for it will turn onto its back and threaten the cat with its teeth. When a large prey such as an adult rat is pursued (which will only be undertaken by a very confident cat that has learnt the necessary technique), the retaliatory threat is much more alarming. A rearing adult rat is a formidable opponent to a cat.

What seems like playing with live prey is an earnest strategy played out between two skilled players. While the raining of paws on to the head of a rat to suppress it is a gladiatorial battle, to most onlookers the same technique used against a small mouse seems hardly fair because of the size difference. Yet it is a fairer match than it appears at first, for a good percentage of prey escapes.

A cat takes its prey back into our homes as its feral cousins take theirs back into the core area of their range. It seems a crazy move, for we are highly likely to say 'Get that thing out of here!', while among a feral group the overlapping area is where other group members are most likely to be

encountered. In either case the hunter stands to lose its prey to others. Yet overriding this is the advantage of taking the prey out of its known area, tightly held in the mouth as first grabbed, and dropping it in the cat's own heartland where it knows the runs like the back of its paw.

On releasing its prey the cat is anticipating manipulating and dazing it fully so that it can be dispatched by the feline neck bite. However, the prey's prime need is to escape. Both are trapped in the survival ploys that best aid their species.

After stunning the prey, the cat cannot just finish it off for it may be 'playing dead' (akinesis), and may turn and bite the cat as it tries to bite the prey. So the cat feigns disinterest and looks about while waiting. This is what the prey has anticipated and makes a dash for it. The cat wheels round and lands a paw again on the fleeing potential food. This is repeated a number of times, but not infrequently the prey reaches some cover and makes good its escape. So this apparent play is a macabre game that both stand a chance of winning.

When a cat brings prey back into the house a number of things are happening. First there is the geographical advantage. Secondly, as in the feral state, the cat is also displaying to other 'cats' that 'this is a good place to be' by its actions. If prey does escape, the cat is only 'stocking its own larder', for a small mammal will normally not regain its own range for a long while. This is one good reason for farmers to have mainly female cats, for males will bring in prey from the fields, whilst most prey catches of queens will have been in and around the barnyard and an escape does not increase the inhabitants!

It can be a real problem when cats bring their prey back to old houses full of hidey-holes, and then lose them. The prey are not usually housemice, so are ill-equipped to survive house-dwelling for long and can die marooned under a floorboard, leaving a tell-tale stench of death.

Another reinforcement to the cat in this behaviour is that we are fairly hopeless 'kittens', showing no signs of catching our own prey. Cats behave towards us in a mixed manner: frequently we are adults and they juvenile, but at other times the reverse, and sometimes the relationship is adult to adult or 'kitten' to 'kitten'.

Cats and the Bird Table
To have a bird table in a garden that is frequented by cats may seem to be providing them with a dining table! But many people like me love to have both cats and birds about, and it is not as crazy as it sounds!

Cats rarely catch birds in contrast to other prey. Both cats and birds are at a superabundance in suburban areas because we provide food and shelter to both. Not having a cat so you *can* have birds, or vice versa, will not change the reality that in your neighbourhood there are huge numbers of both.

I have had a bird table in my garden for years from which birds have successfully fed in numbers, and been helped to overwinter successfully, and all without the avian equivalent of the 'Slaughter of the Innocents'.

It is possible to make your table less user-friendly to cats. Rustic, wooden-legged tables allow the cat to sprint up them, while smooth metal poles do not.

It is also possible to have no pole, and to suspend the table on chains. Poles can have a lampshade-like collar that opens downwards which is effective against cats. Some people like to keep out grey squirrels and pugnacious starlings as well as cats, and so use a roofed table with wire-netting sides or a similar restrictive access system.

I have been able to have co-existing cats by using a simple open table and

Cats catch far fewer birds than they do rodents and other small animals, yet we are often more emotional about their capture

Opposite: In the deadly game of cat-and-mouse

paying attention to its siting. A potentially major problem with an open platform table is that if it is positioned anywhere near a fence or wall, cats can leap on to the table. Placing it in the centre of a vast expanse of lawn with nothing near is no alternative, for you will see very few birds. They need other birds about to convince them that all is well and that it is safe enough to feed. As isolated table will remain nearly bird free, while one adjacent to bushes on which a 'queue' of birds can form will see far more. I found that having my table separated from the fence and a

tree by a japonica (flowering quince) worked very well. It has a most attractive early and long-lasting flowering season which is an asset to any garden, but it also has wicked spiny thorns which cats avoid clambering on, yet which let the birds queue up.

Many of the fears about cats on the bird table are irrational, for in reality cats do not catch birds on the table. Even if they can reach the platform, by the time they are on it the birds have flown. The worst that happens is that the cat just sits on the table for a while.

The real risk is that birds feeding on

much concern. Yet it is due to the privileged relationship which we share with cats that they bring a large proportion of their prey back home, and if they did not, far fewer owners would have cause to worry.

Leroy, one of my cats, is a dedicated hunter, but he is typical in that his prolific catches reflect the richness of prey in the area he uses and the size of that area. I live in a watermill on a river, and there is an extensive rabbit warren a few hundred yards away. During the height of the young rabbit season he brings one home every other day. Yet when we lived in a village, he did not catch one. There, due to a pond in the garden, he brought in many frogs, but rarely killed them.

The availability and abundance of prey makes a big difference. Leroy's catch in the village was much lower than it is at the mill. As a male Leroy had a larger range than my female cat, which gave him access to a wider prey area. While the size of the range of rural feral cats reflects their prey requirements, prey is not necessary for the survival of domestic house cats, their range sizes are independent of its abundance. While this could make them more of a danger to wildlife, this does not occur for a number of reasons.

Firstly, not all house cats are competent hunters, and most only catch prey occasionally. Secondly, city cats catch only a fraction of that caught by rural cats. I discovered this in my work with London's house and feral cats. Although cats are superb hunters, it is their scavenging ability that allows them

It could be a long wait for lunch on the bird table!

the ground are stalked by cats, so it is better to feed them on a table than on the ground, and to try to avoid spillage from the table.

HUNTING

Cats hunt, catch prey, and eat it – they are carnivores. To expect them not to hunt is unreasonable both because of their biology and the natural order of things. Almost incredibly, in the USA there is a growing idea that carnivores are somehow immoral. Although that view may be extreme, that cats catch birds causes cat-owning bird lovers

THE CAT'S EFFECT ON WILDLIFE

Cats have not been found to be endangering wildlife populations on continents (with the possible exception of Australia), but they have had a devastating effect on some island species. A number of island birds became extinct with the aid of the cat, such as the Guadalupe Rufous-sided Towhee, the Diffenbach's Rail and the Sandwich Rail. However, in all but one of the cases such as these, the effect of the cat was part of a package, where intro-duced dogs, cattle, goats, rats and people, had major involve-ment. Only in the most particular case of the Stephen's Island Wren near New Zealand has a cat almost unambiguously pushed a species into extinction when the lighthouse-keeper's cat on the tiny island both discov-ered the species and took its catches back home. (However, even there, collectors were also involved in the bird's extinction.)

to survive as feral-living animals and live with us eating food off a saucer. The urban feral cat seeks both 'waste' food from plastic sacks and bins, and receives food on a plate from feral feeders.

Thirdly, contrary to common belief, cats do not catch many birds, but mainly small mammals. Proportionately, town cats will catch more birds than their country cousins. What is often overlooked is that although cats are far more common in towns than in the country, so are birds! As well as feeding cats, we also feed birds. We provide artificial nest sites in the form of nestboxes and buildings. Our gardens provide good habitat in the form of rich scrubland, with excellent insect support due to an increased flowering time in the year, and lawns with abundant earthworms. Our actions can be seen as providing optimum conditions to maximise bird numbers! Consequently, when Chris Mead of the British Trust for Ornithology assessed the numbers of ringed garden birds caught by cats, he found that they were not having a harmful effect on bird populations.

Although many studies on what cats catch and eat have been made around the world, one by Bedfordshire school teacher Peter Churcher in Felmersham village caught international attention. Unfortunately others made unreasonable interpretations from it, suggesting a 'national carnage caused by cats devouring over seventy million individual prey'!

The error arose by multiplying simplistically the annual catch of the average village cat, which was fourteen prey, by the total number of cats in the country. As most cats live in towns near people, in small ranges and with a fraction of the hunting success of rural village cats, the sum was meaningless. What is significant is whether the prey populations can sustain the predation. On most continents, prey species have survived the attentions of cats for centuries. In Britain, both in Felmersham and nationally, cats only cull 5-6 per cent of the annual productivity of the House Sparrow – which is why House Sparrows continue to thrive.

The Effect of Prey on Predator

Although it might seem that all a prey species does for a predator is provide lunch, they do have a particularly linked relationship; but, who controls whom? It might seem a strange question to ask, for historically human hunters just accepted that predators control prey numbers. However, in 1967, Paul Errington, who had written about the hunting of his cat in Wisconsin, suggested that in good habitat there is a

A cat among the Pigeons in Venice

Opposite: Extinct Stephen's Island Wrens; catalogued and cat caught

'doomed surplus' whose loss does not affect prey density. It became realised that the number of predators was limited by numbers of prey, so ironically prey controlled predators.

I believe that in reality it works both ways, depending on the circumstances. The particular significance of the cat in the equation is that although the completely independent feral cat living on an uninhabited island entirely on its catches is controlled by the interplay between these effects, other cats can step aside with different rules.

The house cat is completely supported by alternative food, so potentially would not be stopped from hunting a particular prey species out of existence locally. The urban feral cat's mixture of scavenged food and hand-outs similarly supports it. Potentially this could make the cat very damaging if it were not for the prey species such as song birds also being supported.

What may at first sight seem simple is further muddied by the scavengeable food available to other predator species such as foxes. Additionally most predators have enough flexibility in their choice of prey species to shift to another when the numbers of one fall, so in that sense other predators are also supported. Consequently we should not leap to conclusions about the damage

Some towns in Australia have introduced 'cat curfews' where cats are taken in for the night. However, the number of native animals caught by cats in urban and suburban Australia is minimal as most have been displaced by habitat change

to wildlife populations by the cat.

That cats can undeniably control prey populations was found during World War II by Elton, who for national survival was investigating ways of reducing rat damage to stored grain in farms. Cats usually have difficulty in killing adult rats, but are expert in catching youngsters. So if the farmer could reduce the adult rat population around his buildings the resident farm cats were able to maintain that level by culling the youngsters.

Strangely, very little research has been carried out on the effect of predators competing with other predators, so the effect of the cat on kestrels, for instance, is unknown.

Cats in rural areas with large ranges are at lower densities, so make less impact on the local wildlife. As Fitzgerald researching in New Zealand has pointed out, at low densities cats may be having little control on rats that may themselves be damaging ground-dwelling species.

The Australian Dimension

The first British cats to arrive in Australia landed with the convict fleets at the end of the eighteenth century. By a century later, another introduced animal, the rabbit, had reached major pest proportions and a widespread policy developed on farmland of breeding and releasing cats in an attempt at biological control of the rabbits. To an extent it worked, for in Australia cats eat more rabbits as part of their diet than anywhere else in the world. But the policy boomeranged as the special marsupial wildlife of the continent, previously isolated from placental mammalian carnivores, began to succumb.

A passionate debate has developed in Australia over cats, and although the cat *is* eating *some* native animals, it may be a scapegoat, for it is only doing so in the Bush, as indigenous animals have been largely eradicated by farming in agricultural areas. Nonetheless, despite fierce opposition in 1995, South Australia passed a Draconian state bill that *any* unidentified cats could be destroyed.

An American Civil War

A similar dispute has been raging in the largest cat owning society in the world, the USA. There are now more cats than any other pet with over 55 million as pets, and there could be around 60 million feral cats. During 1993 and 1994 an alarming battle raged in the country over attempts to have a 'Feline Fix Bill' proposed in the California State Legislative Assembly, in an attempt to reduce feral cat numbers on the assumption that cats were having a devastating impact on American birdlife.

Yet it is too easy to blame the cat just because it is a carnivore, when other factors can be far more significant. Recent attempts in the USA to encourage the neutering of feral cats but to make it illegal to return them to site has demonstrated a lack of understanding of the effectiveness and advantages of neutering and returning them in producing a lower feral cat population. If cats are annihilating their prey species, the controls should be ones that work. Removing cats from where there was once a good population just buys a temporary respite until a new population develops.

A hunting feral cat in Washington DC, USA

Avoiding behaviour problems

Looking wistfully outside.
Housebound cats are more likely to
suffer behaviour problems than those
with access to the outside world

There are three main 'problems' that
cause cat owners anxiety: fouling,
aggression and damage to furniture.
Most 'behaviour problems' are not
problems in themselves, it is where they
occur that makes them problems – in
the home. It is often just normal
behaviour in an inappropriate place –
for us.

The four main causes of the
behavioural problems can be grouped
as: confinement; stress; physical; early
conditioning. Confinement and stress
problems are the major ones.

Confinement
The single main reason for the great
increase in 'behavioural problems' over
the last few years has been the
increased number of cats that are kept
confined. With the cat's range being
reduced to the size of the home,
everything has to happen there, so the
events that would normally be diffused
over a larger area outside are focused
on you and your home. Once cats are
allowed outside to set their own
territorial and range limits, most of the
confinement problems disappear. The
link between behavioural problems and
confinement is something that zoo
managers have to address with regard
to their caged animals.

Confining the independent cat with a
heritage of 3,500 years in the outside
world is bound to cause problems. We
may feel 'safer' inside our homes away
from the apparent risks of vehicles and
disease, but we must be careful not to
transpose our fears onto our cats when
it is not necessary. However, if you live
in a high-rise building in the middle of
a busy city, realistically risks are higher
than elsewhere. If you intend to keep a
cat in a confined area, fortunately there
are ways of circumventing problem
situations.

With our Domestic Cats, readily
available food has already reduced their
need for large ranges. But by confining

them, you are setting the range limits, so individual features of their reduced landscape become more important, and you in particular become the focus for play, hunting and aggression. Not the least of the problems is that a trapped cat becomes a bored cat!

Although cats will adjust with time, I have found that if normally active indoor/outdoor cats are confined, over the first two weeks they become more overt and demanding. If more than one cat is confined, then the frequency of ambushes of one on another increases as the days go by.

If you live in a flat or apartment two or three floors up from a garden, do not automatically assume that you cannot have an indoor/outdoor cat. In Amsterdam, the back walls of many apartments have cat-ramps of single planks with grip slats that enable the cats to go up a couple of flights. At my own watermill home my cats enter the building through a cat flap a couple of floors up!

Stress

A lot of neurotic behaviour in a cat can be caused by stress, and that is often linked to density. Too many cats in a confined space can cause problems. This is often encountered by those who take far too many stray cats into their home to 'save' them from the outside world. The pungent smell of cat spray and urine can often be overpowering in such a house.

Even if cats are unconfined, the introduction of yet more cats can cause a range of problems. As cats recognise our place in their world, so the arrival of a new baby, new partner or even a new neighbour and their pets can cause them stress. For a cat not used to socialising the sudden arrival of a lot of visitors at your home, or the

High cat densities can lead to stress problems, and aggression can become noticeable around mealtimes

incarceration of the cat at the vet or cattery can also be stressful.

In response to stress, fouling can become a serious problem, and aggression can occur. The cat may begin overgrooming. It can undergo appetite changes, losing interest in food or overeating. It can also enter an anxious, depressed state where it may remain for long periods, hunched up with wide-eyed pupils.

EARLY CONDITIONING

The Antisocial Cat

If you have an antisocial cat who does not like people, it probably did not have enough human contact when it was a kitten. But if a social adult cat becomes antisocial, look for a reason for its change of behaviour. It might be suffering aggressive attacks from a nearby dominant tom, it may have been left for a while at the vet for treatment, or it may have been hit by a vehicle. These are the type of events that can cause such a behavioural change.

The Socially Dependent Cat

If your cat has focused its attention on you alone since it was a kitten this can cause particular problems if it is put into a cattery when you go away. Deprived of you and its home, it may become depressed, refuse to eat and lose weight.

Don't Treat a Cat like a Dog!

Training a dog is 'relatively' easy, but training a cat is totally different. The dog with its open geography, hierarchical background operating within a group to hunt, is programmed to co-operate and obey a pack leader. The pecking order is sorted out before the hunt and established by aggression within dog society. We can fit into the niche of the alpha dog, and order a dog what to do.

When a cat hunts it does so alone, so the co-operation via aggression does not occur, and therefore ordered instructions to a cat will have less significance to the animal. When cats do come together and form a group in a semi-social way, it does not revolve around the hunt, and the interactions can be categorised as more affectionate than aggressive. Consequently, in your role as a 'type of cat' in your 'cat group', affectionate interactions will increase your reciprocal bonding with your pet. The simple aggressive dominance effective in dog training is therefore inappropriate for cats.

Although it is possible to establish relative hierarchies among a group of cats, it is less significant and of a lower key than in the dog world. It is possible to view an 'acknowledgement hierarchy' in which the affectionate move of head rubbing may be initiated by a 'lower ranking' animal towards a 'higher ranking' cat more frequently than to a lower ranking one. The parallels are the 'fawning' appeasement moves of dogs.

Once you have trained your dog to carry out a command it will continue to follow your instructions. You can tell a cat to 'get down off that table' and to an extent it may do as you wish while you are around, but not necessarily when you have gone!

Active and Passive Deterrents

Passive deterrents work better in modifying your cat's behaviour than active ones. You can fall back on active deterrents when necessary, but to avoid weakening your bond with your cat it is necessary that the cat does not realise that you initiated them.

One of the simplest active deterrents is a ball of paper that you can lob at your cat if it is making mischief, but it must not be seen to come from you. It has the decided disadvantage that your cat may think it a game and chase after it!

An old standby is the water pistol which can be used when the cat is carrying out the act you wish to prevent. But again, it is vital that you are not seen to aim it at your cat, the water must seem to appear out of the blue. As a technique it should not be overused.

A method employed by animal behaviourists for certain situations uses an upside-down mousetrap. For example, a cat that repeatedly digs up a large pot-plant can be dissuaded by this method. The trap mechanism is set and placed carefully upside down on the pot soil and covered slightly with a piece of paper and soil. When the cat starts to disturb the soil, the trap makes a noise. I am very wary of this method, for cats' paw bones are small and easily damaged. If you do employ this method, take extra care and only use it as a last resort.

Passive techniques are usually easier, both on your cat's and on your nerves, and do not jeopardise your relationship.

The inverted mousetrap for large house plants can be replaced by putting a handful of mothballs in a pierced bag on the soil. Cats dislike the smell – but so might your visitors!

The key passive techniques are best illustrated as alternative approaches to the main problem voiced by cat owners – fouling.

Avoiding the Mess by Passive Deterrents

Once a cat has started to urinate or defecate in your home, it will tend to return and use the same spot. It will detect a residual scent even after a reasonable clean up, and it will seem like a safe latrine. So you need to be thorough. Among the range of disinfectants and bleaches available, avoid coal or wood-tar products such as Jeyes Fluid or Dettol, as these are recorded as having toxicity for cats. Similarly avoid using ammonia-based

If a water pistol is used as an active deterrent, it is vital that the cat does not see its use

109

products, as they will encourage urination because of their smell.

The recommended agents for areas used by cats are based on sodium hypochlorite, and include Milton and Domestos. (Milton should be diluted one part to six parts water, while the stronger Domestos should be diluted one part to fifty parts water. Fully rinse after ten minutes.) Check whether such treatment will damage fabric by trying a small hidden area first. Enzyme-containing biological detergents are also effective and some are sold specifically for this problem.

'Changing the geography' so there is no room for a cat to re-use its old latrine site

Once the area has been made odour-free, you can change the geography of the room. You can put a group of ornaments or even empty milk bottles in the area so there is not room for the cat to squat down and use it. An alternative is to unroll some aluminium foil and place it on the spot. Your cat will not like walking on this and will avoid the location.

Fouling the Floor – Practical Considerations

While stress factors like increasing the density of cats in a home can cause fouling, a number of other reasons within a household can also give rise to this problem.

The cat is the most fastidious of animals. No other domestic animal buries its faeces. If a cat is confined and for some reason your cat does not like the litter tray, then you will find mess around your home. Usually it is the owner's fault rather than the cat's, because of a lack of litter cleanliness.

If a cat starts fouling, your first consideration should be whether you are changing the litter frequently enough. It is pointless to tell a cat off when you are leaving a litter tray uncleaned for two or three days. Cats vary a great deal over this. My cat Tabitha would much prefer to use a litter tray rather than the outside world. So when circumstances mean she has to use one, I could be slapdash and leave it for a number of days without it bothering her too much. Leroy on the other hand, much prefers the garden and is so fussy that he will dig a number of holes before he is happy with the spot. He finds using a litter tray an indignity, so the thought of using even a lightly soiled one is an abomination.

A cat may also foul if it is trapped in for a length of time. It does not have to be physically trapped for problems to occur. For a shy cat, noisy guests can make it feel trapped. A large increase in the number of local cats can cause a shy cat problems about 'going outside'.

Other things can also 'shut your cat in'. When a cat is elderly or infirm, very cold weather can make it think twice about going outside. For some cats, a thick layer of snow can blanket off the soil completely, and you may need to clear a patch for them or use a litter tray at such times to avoid problems indoors. During hot weather female cats with small ranges and a limited variety of places to dig may find the soil rock solid. They would welcome an occasional turning of an area for use! In itself, this is usually not much of a problem, but when a particular stress problem exists, it may tip the balance.

Health problems can cause a cat to foul. A cat that suddenly starts to urinate around the house may not have a behaviour problem but cystitis. Elderly cats may have not only arthritis but also a weakened sphincter muscle of the bladder, and may not be able to reach the litter tray or the outside world in time. In this situation it is pointless to become cross with your pet.

Cold weather reduces the size of a house cat's outside range – but a frosted layer of snow can dissuade some cats from finding somewhere outside to use as a latrine

Fortunately kittens take readily to using litter trays

The Litter Tray

Cats rarely need initial training to use a litter tray, but if you do need to 'retrain' your cat, it is usually a fairly simple procedure of putting your cat in a small room, preferably with floor tiles, such as a bathroom, and introduce it to the litter.

At one time the only choices for litter were earth or sand, but now there is a wide range of commercially available litters with different characteristics and one at least should suit your cat. Unfortunately, your cat may not like the one you find easiest to carry!

Some litters are made from wood fibre, some from corn husk and some from clay. Behaviourally, I find the best litters are those that clump together on absorbing the urine, and dry the faeces, for this makes the excreta easily removable on a daily basis. The litter will therefore be more acceptable to the cat as the odour is reduced. Non-clumping litter will usually need more frequent changing, or the use of a litter freshener – and over-use of that can be offputting to some cats. Multicat households can present problems as some cats find the use by the other cats of the litter unnerving.

For cats that like a quiet corner, covered trays can be an advantage. For owners they also mean that litter is less likely to be scattered around the room and that odours are more contained. Some covered trays even have flaps and air filters.

Litter provides the owner of the indoor/outdoor cat with an occasional convenience when circumstances arise, and also has allowed for the existence of the American-led fad of keeping cats confined. In the USA cat-litter has become a mega-dollar industry with a vested interest in promoting the concept of the indoor cat. The downsides are the behavioural problems of the confined cat, plus the unnecessary ecological disturbances of the industry. The volume of mineral extraction alone is colossal. There is also a real health risk of toxoplasmosis damage to unborn babies if pregnant women come into contact with soiled cat litter.

Spraying

Spraying around the house by cats should not be confused with urination. If a cat is leaving dampness on the floor within a corner of a room it will be from urination, while if it has been left some way up against a jutting out corner of a wall or piece of furniture, then it will be from spraying.

The intact tom will reverse against an object, and with a juddering, vertical tail will release a pungent spray which sends a powerful scent message. The neutered tom or queen can spray, but the event will be of a lower key and the cat will usually position itself less dramatically nearer to a urination stance. Under stress a neutered tom can occasionally spray like an intact tom, but the liquid will lack the smell. The spraying of an intact tom has a territorial marking function, so to assert its right to be somewhere, it will repeatedly spray small but pungent volumes as it patrols.

Neutering is the simple answer for your cat if it is a spraying tom, and after about a month it should cease to be a problem in 90-95 per cent of cases. Behavioural deterrents may be persuasive in the handful of cases that neutering does not control. If they do not work, then discuss with your vet the possibility of hormonal control.

If you are troubled by a prowling neighbourhood intact tom that comes into your home and sprays, then change your cat flap to one with selective 'keys' that are worn on your cat's collar.

Damaging Furniture

A major problem for owners of housebound cats is the damaging of furniture. Even cats allowed access to the outside world will damage furniture, although not as much. Cats need to scratch for a number of reasons, and your best defence is to provide alternatives. A rush mat on the floor will be readily used by some cats but not touched by others. In the outside world, the most frequently used surface is the bark of a favoured tree, so the front of the arms of armchairs are commonly used by cats as 'trees'. The cat will tend to use the same spot time after time, for one of the functions of clawing a tree is to leave a visual territorial mark. Consequently, to yell at a cat after you have allowed it to develop a regular scratching post is unlikely to be effective, as the cat is drawn to the

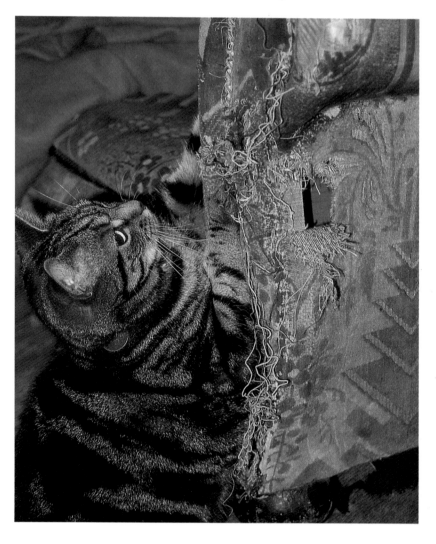

Once a cat has begun to cause damage to furniture, the 'marking' it has made will attract it time and again to what it will come to consider as 'its spot', and its repeated actions can cause severe damage

The placing of alternative scratching 'posts' such as rush matting will help to prevent further damage to furniture around the house

covering the arms with other material or aluminium kitchen foil.

Then, you need to redirect your cat's action onto a scratching post. Many of these are commercially available and are constructed with sisal string, carpet or a similar surface on a wooden post, although they are easy to make. If you have a cat from kittenhood, you should encourage it to use a scratching post. However, if a cat starts to damage a chair or table-leg then place the post directly in front of the position. Transfer your cat across onto the post when it attempts to claw the furniture. You can gently encourage your cat to use the post.

Some people find the active deterrent of the water pistol useful when their cat is in the act of clawing the furniture – but remember the cat must not see it being fired.

Declawing is a barbaric, disabling mutilation and should not even be thought of as a means of control for this problem.

visible damage. You need to prevent it early on.

If the cat's favoured scratching post is wooden furniture then you can use passive deterrents in the form of a strong-smelling polish, or vinegar (check first that this will not damage the surface), which the cat will not like rubbed on. You could place half an orange alongside the problem spot for the same reason. On padded furniture you can change the geography by

Aggression

Aggression is probably the most common reason after fouling why cat owners seek out the advice of behaviourists. Yet 'aggression' is not a straightforward problem to deal with, for it is a catch-all word used by owners for what are a variety of situations.

Even true aggression is coloured by the context, for both an attacking and a defending cat in an encounter are behaving aggressively but in differing ways. Perhaps the most straightforward aggressiveness is that which I have monitored in kittens around five days old. With eyes yet to open, but with sound perception developed and a keen sense of smell, young kittens in a litter at this helpless stage will open their

new cat is introduced. This can be protracted if a proper introductory procedure is not followed. Remember that young animals will adapt more readily than adults. The introduction of

mouths wide and hiss at the smell of strange cats when their mother is away from the nest. This is an entirely appropriate response to an unknown world, and in understanding aggression we need to consider appropriateness.

Many cat owners are alarmed when their home becomes a battlefield between seriously sparring cats when a

a kitten to a resident adult will be easier than the other way round. Similarly, when new cats move in next door, particularly if one turns out to be a wide ranging tom similar to your own, then as the territorial geography is redrawn there will be aggressive disputes. How you react to your neighbours and their cats will affect

INTRODUCING A NEW CAT

The easiest way is to begin with two kittens, as they will come to behave as if they were littermates. When introducing a new cat into a home with a resident adult cat, make it possible for the resident cat to 'meet' just the scent of the new cat for a while to establish the right of the newcomer to be present. To do this keep the newcomer in one room initially, but when the resident cat is absent allow it into other rooms; then put it back in its own room. In this way the two cats gradually familiarise themselves with each other's scent. When eventually they do meet there will be sparring, but it should be less disastrous than it could have been.

The photograph shows two new cats settling their differences: the one lying down is being typically defensive with ears flattened, while the aggressor moves in with its ears flagging its intentions.

Opposite top: Many owners find it alarming that while they are tickling their cat's tummy, it may suddenly grab hold and rake their hand with its back claws

Opposite below: It is vital to spend time playing with your cat if it is housebound, to prevent it becoming bored

A cat 'gym' provides a place where a cat can satisfy its tree-climbing instincts and generally 'let off steam'. Two cats will find a gym a great place for shared 'kitten' games, even as adults

how your cats relate to the neighbours' cats. The dust should settle with a little time. However, do check your cat more regularly for signs of bites and scratches that may develop into abscesses.

Even in more settled circumstances, in particular in high cat density areas, if your cat is a queen she may suffer the unwelcome attentions of a neighbouring tom. While she associates with you as a member of your group, he believes her to be rightfully within his range of operations. This can lead to problems.

The situation usually occurs with intact toms, and you may be able to persuade your neighbour to neuter their cat. Failing that then the water pistol can be helpful, as can a cat-flap with an individual 'key' to provide a safe bolt-hole for your pet.

For indoor cats, aggression towards you and your other cats can arise from boredom. Usually this takes the form of boisterous 'teenage' games rather than real overt aggression, and can be reduced by increasing play and interest in your cat's life by spending more time with it. Also give it somewhere it can be boisterous, such as a multilevel cat gym where it can be harmlessly 'arborial' and remember to integrate it in play with your cat.

'Attacking the Forearm'

Many owners experience what they perceive as aggressive behaviour suddenly appearing when they have been playing with their cat. The flip from play into this aggression is so sudden that they talk of it as schizophrenic behaviour! What has happened is that they have been playing co-operatively with their cat when suddenly it grabs their wrist with its front paws, perhaps even attempting a bite, and vigorously kicks their forearms with its back legs. This is normal fighting behaviour where raking an opponent's belly with the back claws occurs.

Why it has occurred in this situation is that while the cat is playing happily, you have rolled it over onto its back and attempted to tickle its tummy. From the cat's point of view, having had its senses heightened, it suddenly feels vulnerable. As your hand looms over, it seems like a threatening cat and your pet is triggered into a response. (Part sexual, part predatory elements are also involved.) It is so easy to forget the dramatic size difference between the cat and ourselves, which makes even the most confident of cats slightly wary at such times. The cat may learn to go into such a move when our own movements are more boisterous than appropriate.

What should you do? The first thing is to stop moving and 'play dead', and the cat will stop and lose interest. Don't pull your arm away for this usually increases the cat's actions. You may need to distract the cat with your other hand. Sometimes the use of a firm tone can be helpful, but don't threaten to hit your cat. Don't turn it into a game or the cat will keep repeating it.

Remember, you can anticipate the event and avoid it by simply stopping and standing up. You can restrict your play so that you do not roll your cat over. However, a fully trusting cat will be happy to have its tummy gently stroked.

Breeds and Behaviourists

Understandably, far more dogs are seen by behaviourists than cats, for aggression in a dog commonly causes fear in people and can lead to serious problems such as wounding or even death. The linkage of temperament to breed and type of breed in dogs has been recognised in Britain by legal restrictions on Pit Bull Terriers.

In cats, three or more toms are likely to be seen by a behaviourist for every queen as a result of territorial aggression and spraying. It is notable that half of the cats seen are breed cats, although breed cats are only a small percentage of the cat population. While it may be that breeds are more likely to have problems due to their genes, and it may also be a reflection that owners of breed cats are more likely to seek help, basically it is due to the far greater tendency to keep breed cats confined in the home. Burmese and then Siamese generally lead the problems tables.

Keeping your cat healthy

THE HEALTHY CAT

The behaviour of the cat we love can be changed by illness, overeating, being confined and by its recent experiences. The imprint of its genetics and early conditioning are harder to modify, but the other causes can be addressed more directly. Proper cat care does make for an overall healthy cat.

Feeding Your Cat

When we feed our cats it is a social event when we interact together. This is different to the cat's solitary image, yet feral cats often take food back into their group's core area. Our own cats also regularly bring food back into the home. In that sense we are continually being a ready food source. As we bring back food for our cats, in large part we play the mother role and reinforce their 'kitten' status. In another sense, as a continuing dependable food source, we are like a rabbit warren or rubbish-bins around which cats can structure their ranges.

Small cats catch small prey, and so repeatedly eat small feeds. When food bowls are monitored with cats given free access to them, they eat ten to twenty times per day in small amounts. When we feed them twice a day and they rapidly devour it, it is largely due to the way that we feed them.

That is not the entire story, though, for when cats have the opportunity to catch young rabbits, they will take advantage of the glut while it lasts. Although my cat Leroy may dispatch a two-month-old rabbit at one sitting, he usually eats half of a three-month-old the evening he catches it and the other half in the early hours of the morning, after a rest. This style of feeding on larger prey is not dissimilar to their usual feeding pattern in our households.

The Diet

You often hear 'When I come back it will be as a cat for all they do is eat and sleep!' Yet these two features of the cat's life are intricately linked. It is only because it eats the high protein hunter's diet that it can invest in such a lot of sleep – around eighteen hours a day either dozing or sleeping. This reduces stress and extends the cat's life longer than would be anticipated for an animal

Left: Although cats are carnivores they have a need occasionally to chew and swallow grass. Confined cats should be provided with grass in a pot or they will chew houseplants which may be toxic to them

of its size. But the cost to the kidneys of a high-protein diet is a continual barrage of toxic breakdown products. Kidney failure is a major cause of cat deaths.

The Need for Protein

As carnivores, cats have quite different food requirements to people. Don't ever think of making your cat a vegetarian, even if you are. Apart from the immorality of imposing your whim as a flexible omnivore onto an obligate carnivore, you would also kill your cat.

The cat has a uniquely high requirement for protein as it metabolises both protein and fat rather than carbohydrate to produce energy. Cats need particular amino-acids and fatty acids in their diet.

Cat food manufacturers work to produce a well-balanced diet in their complete products. (This is not always so in the gourmet ranges, so check on the label that it is a complete meal if you intend to feed your cat only on this type of food.) Often new flavours seem

OVERWEIGHT CATS

Our cats evolved as lone hunters that must be in peak condition to survive, so they regulate their food intake. Unfortunately, we are not always helpful, for one in ten cats in the UK, and one in three in the USA is overweight. We easily develop rituals of giving titbits to our cats which we find rewarding as it improves our social bonding. However, this does not usually cause cats to put on weight as much as overfeeding at mealtimes. We love our cats to love us so we buy their favourite brand and we ladle on that extra helping.

Many people who go out to work will leave extra food as a compensatory gesture to cover the guilt they feel for their absence. Far from being a kindness, this can lead to weight problems.

Multicat households are a real cause of uneven eating patterns. Some anxious cats are easily displaced from their food, while others are determined to eat anything in sight. You then have to resort to strategies such as feeding them in different areas. Do not leave the food of a 'snacker' exposed if the household also has a fat 'gorger'!

aimed more at the purchaser's taste preferences than the cat's, with combinations like 'salmon and prawn' when what a cat would prefer is 'mouse in a tin'! Nonetheless the can balance aims at being similar to that of prey – around 75 per cent water, with dry weight measured components of 35 per cent protein to 10–15 per cent fat. Dry cat foods are used as a convenience, but they contain only around 10 per cent fluid. As the cat would normally obtain virtually all of its liquid intake from its food, unless it can compensate by drinking about a third of a pint a day the urine can become too concentrated, crystals form and kidney damage result.

GROOMING YOUR CAT

The cat is the most fastidious of animals. Feral cats and many shorthaired house cats go through their lives without anyone grooming them but themselves, and they nearly always look immaculate.

The cat's tongue is covered with papillae or spines which makes it a very efficient comb for grooming. During grooming, the cat also ingests vitamin D from its coat. However, this does not mean you should not comb a shorthaired cat, but once or twice a week will be quite sufficient. It will enable you to check for fleas and to remove shed hair, so your furniture is less likely to end up looking like a Yak. Combing will also mean your cat ingests less hair and so is less likely to have hair-blocked bowels.

Longhaired cats fall into three groups.

Grooming a cat should become a pleasurable ritual for owner and cat alike, and strengthen your mutual bond. It should never be a battle

First, the Northern Longhairs including the Maine Coon and the Norwegian Forest Cat, have heavy coats. Second are the lighter-coated Angoras, Balinese and Somalis, and thirdly the modern Persians. The natural coats do not need the intense attention that the modern Persian requires.

Breeding has modified the modern Persian's coat to such an extent that it is essential to brush and comb it daily. A modern Persian with a seriously knotted coat is a distressing sight, for the animal is unable even to walk properly as its skin is held taut in patches. This type of cat does not survive in a feral state. The older-style longhair coated cats do not have the problems of the modern Persians. The traditional Angora undergoes a dramatic moult each spring, for its original home of Anatolia is freezing in winter and blisteringly hot in summer.

Modern Persians need gentle grooming from kittenhood so that they do not become agitated, but fortunately most have a good temperament and enjoy the attention. Ease out knots by using a wide-toothed comb or by hand. If the coat has become heavily knotted seek a vet's help.

The buckled tear ducts of flatter faced longhairs cannot drain properly, so fluid wells out of their eyes and matts the fur. Gently wipe with cotton wool, but avoid contact with the eyes.

The coat of longhair cats can become matted around the anus, and can become infested with blowfly maggots in hot weather. Grooming is essential to avoid this.

Fleas are more likely to reside where the cat finds it harder to reach with its probing, rough tongue. An urgent piece of washing usually indicates the presence of fleas; however, it can also be a habitual displacement activity, to cover that it is unsure of its next move

FLEAS AND OTHER PARASITES

Grooming will soon tell you if your cat has fleas, if its scratching or sudden attacks of washing have not already alerted you. Comb your cat on some white paper, and if characteristic black specks appear, you have found flea droppings. If your cat has fleas, so does your house and both will need treating.

Fleas are host specific, but if for some reason a cat is away for some weeks they will have a desperate attempt at feeding on us. The fleas lay their eggs on carpets, bedding or crevices in the floor, as readily as on the cat. Warm summers and central heating boost their numbers.

The Californian approach in a hot climate to cat-flea 'plagues': the electronic collar sound system

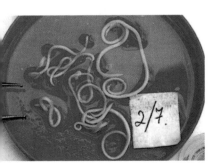

Roundworm

Tapeworm

The vacuum cleaner should normally contain the flea problem around the house, augmented with an antiflea house spray. For the cat, flea collars are helpful, but should not be used on kittens. Examine the neck area a few days after introducing your cat to a flea collar, as the skin of some cats has an inflammatory reaction to them. Insecticidal sprays and powders are effective, but many cats find the hissing from a spray unnerving, so a pump action may be better. Sprays and powders only kill the fleas on the cat at the time. As flea eggs hatch the cat will need retreating. Oral dosing of the cat to control the fleas is becoming popular, as is the treatment involving a drop of a particular insecticide applied at the back of the cat's neck where it cannot reach and is absorbed into the body.

Some cats develop a flea allergy and the animal's repeated washing of the affected area worsens the situation. This may need veterinary attention.

Cats can develop tapeworms from catching prey, but the most common tapeworm, *Dipylidium caninum*, can also be caught by a cat swallowing a flea during grooming. Although the immediate host is normally a rodent, this parasite's eggs can be eaten by the flea larvae and then taken up by the cat.

OTHER COAT AND SKIN PARASITES
Ticks

About 80 per cent of ticks found on an afflicted cat are around the ears. Vets often recommend the application of alcohol or insecticide before removing the parasite. Use tweezers to remove it cleanly, taking great care to pull vertically so that the mouth parts are not left in the skin, otherwise it will form an abscess.

Ear Mites

These minute animals are the most common cause of problems in the cat's external ear, and will spread to other cats and dogs. Kittens are particularly vulnerable and a plug of brown wax caused by a large number of mites can form in the ear. Vets will prescribe ear-drops and treatment is essential, otherwise a secondary infection can cause further damage to the cat's ear. Regular treatment at a time when the young cat needs to develop confidence with its new owner is behaviourally unfortunate.

Harvest Mites ('Chiggers')

The tiny red dots moving around on the cat's legs, feet, head and ears in the autumn are harvest mites that can cause irritation. Use flea spray to kill them.

Fur Mites ('Walking Dandruff')

This is normally spotted when the cat is seen to have dandruff on its back. Fortunately the mites are not common as people can become accidental hosts. The mites can be treated with flea spray.

Lice

Lice are uncommon in healthy cats but they can be found on sickly animals. The eggs ('nits') stick to the fur and look like fine dandruff. They are controlled by flea spray.

INTERNAL PARASITES

An active hunting cat will pick up internal parasites from an intermediate host such as a rodent or bird. While grooming your cat you may find evidence of a tapeworm in its intestines from what seem like grains of rice in the coat. More disturbing to most owners is the appearance of a small, flat, white, moving creature in the fur near the cat's rear end. These are motile segments broken from the end of the tapeworm inside the cat.

The hunting cat can also pick up roundworms. Kittens can be infected during birth. You may find roundworms when your cat brings up a hairball. You should obtain courses of treatment for tapeworm and roundworm from your vet.

Toxoplasmosis

Toxoplasmas gondii is a single-celled organism that has the cat as its primary host. With the growing tendency to use cat litter, its life cycle needs to be understood because of the potential risk to unborn children.

Cats can be infected with the organism by catching prey. We can be infected by eating and handling raw or undercooked meat, by not washing our hands after gardening or playing with dirt. An infected cat will shed oocysts in its faeces. These can be picked up directly by another cat and infect it, or it can be picked up by any bird or mammal – which includes us.

In most people infection may not cause much effect, but in those with reduced immunity it can be serious. For

Below: Health check: vet listening to heartbeat and breathing, then examining mouth and teeth

Bottom: Wrapping a cat in a towel makes it easier to administer a pill by yourself at home

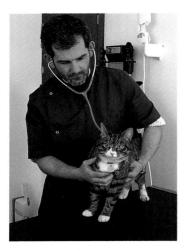

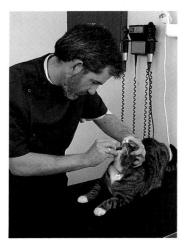

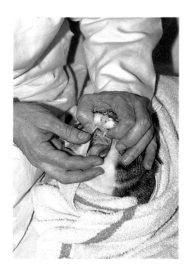

an unborn baby with an undeveloped immunity there is the risk of stillbirth, mental retardation and blindness.

Cats are fastidious in their cleaning so there is little risk of infection from handling them, but it would be safer for pregnant women to avoid contact with soiled cat litter. Use disposable gloves or, better still, get someone else to be responsible for changing the litter. If the litter is changed daily the oocysts will not become infective in that time so the risk is greatly reduced. Outside of warm climates most people are infected by undercooked meat rather than directly from faeces in soil or litter.

CAT EQUIPMENT
Cat-flaps
For a cat to have its own door to a building is not a modern idea. I live in an ancient watermill which has had a cat hole in its door for hundreds of years for the working cats.

Today's cat-flaps are an excellent development, for the flap keeps out cold draughts as well. Unfortunately, the simple flap can allow neighbouring cats access if they know how to use flaps and are determined to enter. The scent of your own cat on the flap is usually enough to inhibit entry. If you are troubled by other cats, there are the additional options of a simple magnetic release or an electronic cat flap. Both allow access only to those cats with a magnet or colour-coded collar tag.

Training for the Cat-flap
Some flaps open only one way and are not transparent, so the cat has to learn

to flip it up and work its way under it Most, though, are two-way, so all the cat has to do is push. If the flap is transparent, the cat can check if it is safe to go through.

Whether the flap is one-way or two-way, initially it should be held wide open either by tape or a clothes-peg high up on the flap's side. Allow your cat to familiarise itself by sniffing the opened cat-flap first. Encouraging the cat to pop through a few times by calling it is usually sufficient. Over a number of days, gradually lower the flap until there is only about a 3-inch gap and the cat has to push the flap to get through.

Once the cat is coping with that, drop the flap completely, and the cat usually manages to open it. If the flap is either magnetic or electronic, the triggered locking device is usually better taped down or switched off so that it can work as a simple flap when the cat is learning. Once the cat is confident about using it, bring the selective device into operation.

Cat Collars
If your cat is an indoor/outdoor cat, it will need to have some form of identification. The simplest is the cat collar with a small name tag with your cat's name, your surname, address and phone number. If your cat gets injured or strays, it is vital that you can be contacted.

A name tag also means that your cat is unlikely to be rounded up as a stray, or mistaken by a not-so-near neighbour who might otherwise assume it was homeless. Toms may have more than

124

VACCINATIONS

Two vaccinations are currently routinely given to cats against Feline Infectious Enteritis (FIE) and 'Cat Flu' respiratory disease (Feline Viral Rhinotracheitis (FVR) and Feline Calicivirus Disease (FCD)). Once the passive immunity received by kittens from their mother's milk fades, it is important that they are treated. Your vet will advise you on timing. For show cats and any cats spending time in a cattery, it is essential that they receive their injections.

In those countries where rabies exists it is advisable for your cat to have inoculations against it.

Cat wearing its kit of name and address tag, night reflective collar and colour-coded 'key' for its magnetic cat-flap

one 'owner', as they range across a number of gardens and may be fed at more than one house.

Collars must have an elasticated part so that if it is snagged, the cat can pull itself free. If it is fitted too slackly, snagging on tree branches becomes more likely. There should be just room for two fingers under the collar. If your cat is likely to walk across a road at night or even dusk, then a reflective collar makes it more likely to be seen.

A very few cats manage to make a habit of catching a tooth on their collar, and for them the injectable identity microchip is an option.

Cat Leads

There will be times in all house cats' lives where the use of a lead and harness is invaluable. It is especially useful after you have moved home and are introducing your cat to the outside world after it has become familiar with the new house. When buying a harness make sure it is big enough.

Many cats initially attempt to wriggle or squirm out of a harness, or just flump straight down onto the ground. As you will not want your cat to run off in a panic in an unknown place, it is vital to allow it to become used to the harness indoors. Let your cat wear the

Injector 'gun' and reader for identity microchip. The chip is lodged in a transparent block only so that it can be seen more clearly for the photograph

Cat doors are not new! Cat using the forerunner of the cat-flap in the door of an historic water-mill, provided to give access to the working cats of long ago

Opposite: Introducing a cat to the garden of its new home by allowing it to investigate on a long lead

harness without the lead for a while at first. Lengthen the time it wears it, then connect the lead but allow it to remain slack and preferably do not hold it at the beginning.

If you drop the lead accidentally when you are with the cat in an unknown or potentially dangerous place, do not shout or charge after the cat for this usually encourages it to run from you. It is far better to walk quietly and gently alongside the cat without looking as if you are about to pounce, and pick up the lead without a fuss.

TRAVELLING WITH YOUR CAT

Cats can become stressed very easily when travelling by car, and may call out repeatedly. They may work themselves into a state and be sick in only a few miles. In hot weather, it is particularly bad for cats to become stressed in cars.

To avoid these problems, it is vital to gently condition your cat to travel, preferably from a young age. First make sure your cat is happy to use a carrier or it will probably struggle as you approach with it in your hand, especially if you only use it when you visit the vet. Ideally the carrier should be a ventilated plastic one which gives the cat a feeling of security, and of not being exposed on all sides. Wicker carriers are not necessarily very secure or easy to clean.

Put in some newspaper, which cats

love to sit on, and gently put your pet in but don't shut the door. Leave the carrier around the house for some days so that the cat can use it as a nest. From time to time, walk the cat around the house in the carrier.

Initially, put the cat in its carrier into the car, and sit with it there. Do this for a few days, then again with the engine idling. Introduce the cat to very short journeys and progressively longer ones. Ensure that you drive smoothly and not too fast, for the first few miles of the journey. Motorways are generally easier for the cat as they have fewer bends, roundabouts and junctions.

For too many cats, the only journeys they make are to the vets or a cattery, which gives negative reinforcing. As on some occasion you will probably have to carry your cat in an injured or seriously ill condition to your vet it is important not to add unnecessary stress because you have not habituated your cat to travel.

Moving Home

Moving house is a very stressful time for us, partly due to the separation and loss of the familiar place, friends and memories, and partly because of the trauma of the upheaval and the arrival of the unfamiliar. It is the same for our cats. We at least know why the upheaval is taking place, but for the cat it is just perturbing.

Do not consider your cat at the last moment, but put it into a quiet room while the rest of the house is being emptied. If the cat has been scared and has run off at your old home, you can

Introduce your cat to a cat carrier so that it becomes familiar with it before you need to use it

sit and await its return. At the new home, no risks should be taken.

Take the cat in your car in its carrier to your new home, and put it into a room where it will not be disturbed. Provide some food, a litter tray and lots of reassurance. If the rest of the house is in a more settled state on the following day, begin to allow the cat to investigate. Do not allow the cat into the outside world for at least a week.

When you do allow your cat out for a brief sortie, it is safer that you take it on a harness and lead. Certainly, ensure that you remain near the cat's entrance back to its new home at first so that it becomes familiar with its safe return route. Do not leave the cat

outside alone, but stay quietly with it. Gradually lengthen the time outside, and you will be able to gauge when it can be allowed to look after itself by its growing confidence.

THE ELDERLY CAT

Cats wear age with great dignity. My own old female cat, 'Mr Jeremy Fisher', familiar to readers of my book *The Wild Life Of The Domestic Cat*, reached almost twenty-five years of age. She remained remarkably fit, and until just a few months before her death, still walked along the top of the garden fence.

Intact toms can expect to live at least two years less than neutered toms. Neutering queens has little effect on their longevity. While twelve to fourteen years is a good age for a neutered cat, male or female, few will live above sixteen. Some, like Jeremy, will make it into their twenties.

As the cat becomes less active, you may need to trim its claws more frequently. Its walking may become stiffer and the coat less lustrous. Nonetheless, it will still enjoy life, particularly snoozing in the sun.

You should pay particular attention to the cat's teeth and gums. If it does not stress the cat, it can be helpful to clean its teeth. If tartar builds up, gums become inflamed (gingivitis). Your vet can remove the tartar, or the teeth may become infected and be lost. Giving the cat some real meat to chew on occasionally can help it to clean its own teeth, unlike canned food.

The most usual cause of weight loss in the elderly cat is partial kidney failure. It may drink more. The cat's high protein diet makes its kidneys particularly vulnerable. Consult your vet to see if this is the cause of your cat's weight loss. Some cats respond to treatment and a modification of diet for a while. It will retain a reasonable quality of life for a while but the point will be reached when this cannot be ensured. You should then discuss with your vet the realistic quality of your cat's life and what it can expect.

Euthanasia

Your pet will have become a loved member of your family and may well be your closest companion, so your grief will be real if it has to be 'put to sleep'. Your vet will appreciate your dilemma and advise you accordingly. But it is your pet, and the final decision will be yours. Discuss with the vet whether the act should take place at your home or in the surgery.

Normally, pet cats are put to sleep with an injection of barbiturate. You will also need to discuss with your vet whether you wish your cat to be buried in your garden, or buried or cremated at a pet cemetery, rather than being disposed of in a more routine way.

Around 3 per cent of all veterinary consultations involve euthanasia. These can be upsetting for the vet as well, despite being routine. Courses in bereavement counselling are becoming normal practice in veterinary schools.

An elderly cat, almost twenty-five years old, basking in the sun. Despite showing weight loss due to advanced kidney damage, it still enjoys snoozing in a favourite spot

Life with our cats

A young professional couple saying 'good-bye' to their cats as they leave for work. Cat owners acquiring a second cat to keep the first company when they are out has been a major reason for a recent massive growth in the number of cats owned in Britain and America

localised islands of populations in and around urban centres, which enhanced a growing genetic diversity augmented by natural and artificial selection. Starting from a small population in the Middle East, cats are now spread worldwide in astronomical numbers that would normally be denied a wild predator.

On the down-side has been the witchcraft era (see pp29–30), current breeding trends which have disadvantaged cats, and stresses caused by the recent trend to confine cats. High suburban cat densities increases the probability of feline fighting and traffic accidents.

Anthropomorphism

Throughout history, our continuing relationship with the cat has been more obviously coloured with anthropomorphism than with any other animal. It had its deified Egyptian ups, and its sixteenth- and seventeenth-century European downs. Habituation was led by French aristocrats, writers and actors. The early years of the cat fancy were assisted by two artists, Harrison Weir and Louis Wain. Showing promoted cats, and they became accepted as pets. But what is a pet? Wain's anthropomorphic portrayals of cats, in which he showed fluffy dressed Persians, blurred the line between cats and people.

Today, many people are uneasy about the desirability of showing cats, for they believe that an animal which prefers territorial confidence should not be displayed alongside others. As with

THE SUCCESS OF THE CAT

Along the domestic route cats have engaged in a trade off in which overall as a species they have done spectacularly well. Those that could initially tolerate being near people and their artefacts had access to a superabundance of food, which enabled their numbers to soar. Their distribution by people around the globe led to large the argument against zoos, it is suggested that television and books keep the cats' positive profile high. Although my TV *Cats* series has now been shown around the world, when made its documentary format for cats was an unusual venture for television. Showing remains a showcase for cats, and the fierce anti-cat lobby in parts of Australia and America demonstrates the need for exhibiting the cat positively. Unfortunately, showing by its very nature promotes the breed cat over the moggie.

In Britain, there are two major cat shows each year – the National and the Supreme. At the latter as well as showing the cats, there is a competition for the best decorated cage. To some, the turning of an exhibition cage into a stage set is a harmless bit of fun and makes the cat's cage less austere. To critics, it seems as if the cage is a doll's house, and not far from making Wain's dressed-up cats a reality.

Cats and Writers

The popularising of the cat by writers and artists such as Edward Lear and Beatrix Potter have made them endearing images for children's pets. Edward Lear was so devoted to his cat Foss that when he moved home, his new villa was constructed exactly as his old house so as not to inconvenience his cat.

Not wishing to inconvenience the cat is common to many writers, though not usually on such a grand scale. The list of writers who wrote with their cat as companion range from the worthy Sir

The blurred reality of anthropomorphism: me and cat, with young women dressed up as cats, 'borrowed' from the stage musical *Cats*, at the launch of my BBC TV series *Cats*

Walter Scott to the scary Edgar Allan Poe, and from the cat-like Colette to the macho Ernest Hemingway. Many, including myself, will write around the cat rather than disturb it. Why do we do this, and what is the basis for the cat's special relationship with writers? Writing is a solitary profession carried out quietly, often at home, making the writer an ideal focus for the cat, while

A cat makes an ideal companion for the retirement years

the cat is an ideal companion for the writer, as it likes to sit and sleep beside you for hours. This behaviour is equivalent to related feral cat group members sitting alongside each other, or cats that sit together on the sofa for hours. These restful, non-threatening sessions reinforce group bonds and assure the occupation of a key part of a range. Yet it becomes a time of apparent communion and empathy to the writer and cat as they are wrapped in contemplation. Small wonder the cat developed powerful friends among artists and writers that turned around the public hatred of cats.

While anthropomorphic attitudes can be derided, they are at the centre of our relationship with cats, as behaviourally in many ways we relate to cats as if they are children. Yet anthropomorphism is a two-way street;

perhaps we need a different term to describe when cats behave towards us as if we are cats (feliomorphism?). We both recognise the other as a different species, but behave as if they are not at the same time. Honorary people and honorary cats!

Cats fulfilling the role of perpetual children are responding to a growing need to bring contentment to an ever-increasing population of lonely, elderly people, and non-demanding companionship to the large numbers of single people and single parent families. Over the last decade cat ownership has increased dramatically by a third in both Britain and the USA. One of the main reasons for the increase is that young professional people are choosing to have cats instead of dogs as they are less restrictive, and are often child-substitutes while they delay having a family. With both partners out at work, they feel that the cat needs a companion, hence the massive increase in two-cat households.

While some adore the cat as a dependent toy, fortunately others still love cats for their free-living spirit – the cat as cat.

Changing Attitudes to Cats

Unfortunately for American cats the USA is leading the world in making the cat a dependent, trapped toy. After centuries of no-nonsense American attitudes toward animals, ideas on feline behaviour in the States are changing for the worse. Fifteen years ago most cats in the USA, like Britain today, were indoor/outdoor cats. Absurdly a creeping fear of the outside world as a place for pets has gripped America. The hazards for cats are seen to be primarily traffic death and disease risk. In Baltimore it was found that a cat had a one in ten chance of being killed in a year on the streets. Understandably where traffic is dense there is a reason to be cautious. Disease risk figures are often expressed in the USA as if preventative injection courses did not exist. Rabies is endemic in the USA, but the main risk of the disease is from dogs. Cats rarely communicate rabies.

Before keeping a cat confined you need to weigh up the possibility of disease risk and road death with the probability of confinement behavioural problems. Now the USA has a behavioural disaster as most of the 55 million owned house cats are confined. Urinating around the house,

'clawing the drapes' and aggression are now significant parts of the American cat's lifestyle.

Alarmingly, at a recent meeting of American cat welfare organisations, one senior executive said, 'Now I assume that one thing we can all agree about is that all cats should be kept in all the time?' How has this happened in 'the land of the free'? Influential breeders

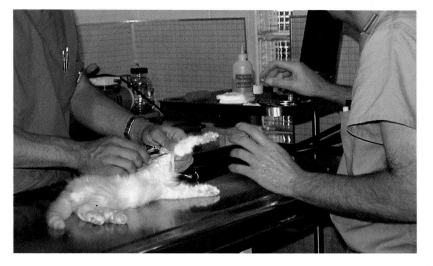

A kitten undergoing a declawing operation in the USA

and writers such as Milan Greer promoted the idea of confinement in the 1960s. Ironically, the old natural breed of which America is rightly proud, the Maine Coon, developed as a robust cat because of its outdoor barn life.

Confined cats cause 'problems', and owners often seek drastic remedies. Although declawing is justified as being the last ditch stand against 'aggressive' or furniture-damaging cats, it can be the first resort and carried out by vets on kittens. Although a caring and original writer, Greer recommended declawing, but he did at least argue that cats should be given the benefit of the

You can pour out your heart to your cat – and you can trust that your secrets are safe. A boy with his cat in New England

doubt up to eight months of age. This barbaric, unnecessary operation has become common in the USA. This mutilation of millions of cats is a real cost that needs considering when policy statements are made by welfare groups about captive cats. Fortunately, the British Veterinary Association and the Royal College of Veterinary Surgeons are firmly against declawing, and consequently it is not allowed in Britain.

Behavioural approaches are far better, and today practising cat behaviourists are often available for consultation via veterinary referral.

The Temperament in the Cat–Human Relationship

Obtaining a cat 'just happens' for most of us! A fifth of the population of house cats walked in off the streets! Some are given pets, others inherit theirs when family members move or die. The majority seek out their cat as a kitten, either from someone with a litter or from somewhere like a Cats' Protection League shelter. The CPL, the largest charity in the UK concerned solely with cat welfare, annually finds homes for

over 75,000 cats, but handles more than double that number each year. A minority of those seeking cats want particular breeds and so visit a breeder.

Cats have individual characteristics, and one type will suit you better than another. Of breed cats, if you want primarily a lap cat, Persians tend to be more placid, while oriental cats, particularly Siamese, are outgoing. If you are choosing a moggy look more carefully at the individual. When obtaining a kitten select carefully. Remember this relationship may last longer than your marriage – so give it more than a few moments thought! Intriguingly, the feature that people normally use to select cats, regardless of behaviour, is colour.

Check that the kittens have been habituated to people. When making your choice, see how they respond to you. Appearance can matter, but so can behaviour. Kittens played with by a number of people become less scared of people generally. Early handling habituation would have been an enabler during the early period of domestication when little genetic selection towards tractability would have occurred.

It is not unusual that a cat has become primarily attached to one person but has remained shy of other people. Do not despair, the cat can be changed by greater habituation to people in adult life and it will improve, but its dye is cast.

A cat's behaviour can be modified, even though the sensitive period is long past. You can increase the bond with your pet. Don't just feed your cats, but spend time with them, so each day you can enhance the link.

Cat Communication

Our relationship with cats is two-way. As we are two different species it is remarkable that we can communicate at all. Our historical relationship with dogs is instructional as they are a pack animal and we can elicit a trained response. In contrast the independent cat does not usually play that game! However, in our shared lives the mutual responsiveness between people and cats is based on certain understanding and expressions of intent that are communicable.

Mildred Moelk in 1944 distinguished sixteen utterances in the adult cat which she found to impart specific meaning.

The 'song' of caterwauling is made by two facing cats, usually males making sabre-rattling threats. Due to the high density of cats in suburban areas, town dwellers will have their sleep disturbed by caterwauling far more than in rural areas.

The shrieks, screams and snarls of the active fight are also unambiguous. Cats also have a melancholy distress howl. The distinguishing vocal feature between the small cats and the big cats is that small cats scream while big cats roar. This has been believed to be due to the hyoid cartilage being bony in small cats and flexibly cartilaginous in the big cats. This has recently been challenged by the discovery that the big cats have a large pad of elastic tissue connected to the upper part of the vocal cords allowing the roar, while it is missing in the small cats.

Most of the cats' vocalisations that we hear are lower key. Some cats are very insistent with their request 'meows' towards feeding time, while others remain mute. When cats greet each other or us, they will make a rising trill as they arch their back a little and bounce lightly on their front paws. As we walk towards the house they will greet us in this manner, or rise up to rub their head or mid-back against our hands.

Cats appreciate that our world is vocal, and turn up the volume in comparison to the sounds made among feral cats. There has been positive feedback. They find that when they call we pay attention. Similarly, they sit in front of a door, and we let them out. The communication is certainly two-way, so we do train each other into repeated patterns of behaviour. We both reinforce our approval, for we stroke the cat's back, and it rubs around our legs.

We seem like large mother cats to our pets, and they purr contentedly. Sitting together we relax, and our stresses fade away

Cat communication between mother and kitten is gentle in contact and sound, and purring bonds them perfectly

Purring

The most cat-like of all sounds is the purr – it is in essence feline. It is at the heart of our close relationship with the cat. Despite this there are very few early historical references to purring in contrast to other feline features such as the shining eyes. One does occur in the eulogy to Beland written by Joachim du Bellay in the mid sixteenth century:

> *He was my favourite plaything*
> *And not forever purring*
> *A long and timeless*
> *Grumbling litany.*

When purring is written about by authors, it gives an indication of the closeness of their relationship as it generally occurs at times of affection. The fear of cats that stalked peoples' minds for centuries, putting the cat in league with the devil, prohibited any intimacy with them. Cats were kept to the utilitarian task of rodent control of barnyards. Clare Necker drew attention to the lack of an ancient Greek or Latin word for 'purr', which is consistent with our understanding that they did not keep cats as close pets.

Remarkably, there is still debate as to how the purr originates. One idea was that blood surging through the inferior vena cava produced a sound as it forced its way past the central tendon of the diaphragm. It is now generally agreed that, as with most vocalisations, the purr originates in the larynx. Muscles cause the vocal cords to close until air pressure forces them open. The involvement of the false vocal cords seems probable as vets can avoid the problem of purring drowning out all other sounds on their stethoscopes by gently squeezing the larynx, which inhibits the false vocal cords while leaving the inner vocal cords unaffected. The diaphragm is also involved, as its rhythmic tensioning is essential for air control of the larynx. Indeed, purring is the 'purrfect' cure for hiccups! Hiccups are due to involuntary spasms of the diaphragm and the tensioning of the diaphragm muscle during purring stops that.

A cat on our lap being stroked purrs in contentment. Yet before the close relationship of domestic cat and mankind, small felids were making the sound. It is the singularity of purring that makes it such a significant sound. It fulfils a role that is important to the cat family and particularly to the small cats.

To seek why our cats purr it is necessary to ask why most other animals do not. What is it about the cat's way of life that makes it so vital?

However, purring is not completely exclusive to cats. Hyenas and Viverrids are the closest relatives of the cat family, grouped together as the Feloidae, the cat-like animals, and some make a sort of purr on suckling. Generally, across the three families, the animals tend to be solitary hunters and nocturnal.

When your cat purrs and kneads upon your lap, it is behaving towards you as if you are its mother. This fact was noted by Charles Darwin in 1872 in *The Expression of Emotions*:

Now it is very common with young cats, when comfortably lying on a warm shawl, to pound it quietly and alternately with their fore feet; their toes being spread out and claws slightly protruded, precisely as when suckling their mother. That it is the same movement is clearly shown by their often at the same time taking a bit of the shawl into their mouths and suckling it; generally closing their eyes and purring from delight.

Leyhausen believed that the function of purring was a signal from the suckling young to their mother that all was well. Once in the feline repertoire, a mother was able to use it to reassure the young during suckling. However, many animals suckle but do not purr. In

During suckling, both mother and kitten will purr

A defensive, hissing snarl. In the defensive cat the pupils open wide, while those of the aggressor are in a tight slit

Opposite top: The extreme arched position of the young Agouti cat in Thailand combines features of both a defensive posture, and the one which threatens attack. Inexperienced youngsters will adopt it as a 'safeguard' pose which they can develop either way

Opposite below: The role of aggressor and defender is often more fluid than it seems, but here the closed pupils give a certain menace to the fistful of flick-knives

kittens it starts when they are about a week old. Before that age their limited movements do not seem to require the mother to hear the signal. Unlike sounds using the true vocal cords, the purring arising from the false vocal cords can be made with the kitten's mouth shut firmly clamped around a nipple. The reassurance purring by mother and littermates reduces movement and keeps the young in the nest.

The sharp, protractable claws of the kittens may be one reason for the purring. In settling the young down quickly, the mother is less likely to have her teats lacerated by scrambling kittens. Viverrids also share with cats the otherwise unique protractable claws.

Males provide a huge territorial area within which females range. However, males do not provide food, so the female rears her young alone and hunts on a solitary basis. Purring between the young kittens while she is away from the nest will keep them together. When she returns to feed them, the mutual purring sustains their quiet closeness.

Much of the mother's time is taken up by nursing the kittens in a hostile world. It is vital that some *quiet* continuous signal can be used to inhibit kitten movement and noise.

The purr is also a statement from the kitten that says 'I am here, I am here...' It is this statement of existence in the purr that is noticeable when a cat is snoozing unseen under a duvet and you sit on the bed. Suddenly up comes a purr, not of contentment or greeting, but a statement that 'I am here'. It is in its anxious form as an expression of existence that the vet often encounters a cat purring on being examined.

When cats purr on our laps, most of the behaviour, as Darwin appreciated, is neotonous as the cats are made to feel like kittens by our large, maternal-sized, warm lap. We do the equivalent of licking them by stroking them with our hands. We are reinforcing our bond, for a cat family that purrs together, stays together!

Cat Etiquette

If you have difficulty 'making friends' with strange cats, you can be placed in the awkward situation of visiting friends who say 'Oh you like cats, don't you? You'll love Binny', and Binny disappears!

We can be so keen, yet simultaneously anxious, that the cat is alarmed. Remember that communication is two-way. Ensure that you are not threatening, and acknowledge cat politeness.

Don't charge in at your full height – squat down. Initially, extend your arm

138

low down, with your index finger only stretched towards, but not reaching, the cat. Once the cat has accepted that, advance your finger so the cat brings its nose forward to investigate. It will then usually recognise your finger as the 'nose' of your hand which is cat-head size, and simultaneously as a finger that it will use like a scent-stick and rub its mouth against its point. Only then tickle its head and even shy cats will usually accept you, and your friends' response will be one of surprise: 'Oh Binny doesn't usually like strangers, she can tell you're a real cat person!' Your credibility will have shot up!

In a room of strangers, cats often go to the person who is scared of cats, and sit on their lap. This perversity is due to

their perception that the stock-still person will be less inclined to hassle them. Consequently, do not overdo the stroking of cats on your first meeting, because they are sizing you up.

There are also ways of relaxing cats at

Many owners recognise the stress relief they gain from cat companionship in a troubled world

reassurance. Your cat will respond in like manner – direct communication! Similarly, a long yawn is a useful addition. With all cat communication, there is the appropriateness of doing the right thing at the right time.

Facial Expressions and Body Posture

The cat's range of expressions is not limited to vocalisation. A whole range of facial and body positions have specific meanings and the cat's ability to mix them allows it a wide spectrum which is superimposed upon the vocal component. Paul Leyhausen established a gradation of both body postures and facial expressions that show a recognisable span from fully defensive positions through a neutral position to full threat-of-attack positions.

In extreme defensive mode, the cat's ears are flattened to its head so that they almost disappear. In contrast, when a cat is aggressively moving towards a confrontation, its threat to attack can be read from the oblique angle of its body in the move, its head held down slightly, and its ears drawn round so that the backs can be seen from the front. The aggressor initiating a fray will often be poised to lift up its front paws onto its opponent. As it moves forward, its rear will be angled higher than its front. The cat anticipating such an attack, as well as flattening its ears, will also usually lower its head and shoulders which will allow it to take a floor dive where it can defend itself by raking its hind legs back against the expected launched attack. The relative balance between aggressor and

a distance that I have established from working with feral cats for years, and they work just as well with house cats. A long, hooded, leisurely blink is very calming, and used reciprocally between cats when they are settling as mutual defender can fluctuate during both the build-up and the combat if it occurs.

Mock battles can break out in quite vigorous 'games' between adult cats. They can become ritualised, in that unexpectedly the cat initiating the sequence may be the 'defender', which induces the 'attacker' to take part, and the 'battle' can become an excuse for a race round the house.

Stroke That Stress Away!

Cats are good for you! Stroke a cat and your stress levels drop. The people who need cats most are doctors, for they have around the highest suicide, divorce and stress levels of any profession. Physician heal thyself – get a cat!

For some years research has shown that pet ownership is good for your health and the study of 'companion animals' has become respectable.

The Cats Protection League surveyed owners and found that the reduction of stress, and the relief from depression and anxiety were major perceived advantages of living with a cat. Many long-term psychiatric hospitals and old people's homes have had resident or stray cats whose company has been enjoyed. Dennis is the 'official cat' at the Luciel Van Geest Centre in Peterborough, where Staff Nurse Pamela Baines finds that certain patients put significant effort into his care. She also finds that he 'certainly helps to relax and calm stressed or agitated' patients. The Cats Protection League discovered that 'eight out of ten caring organisations found that cats helped people get better'.

Pets help institutionalised or isolated people feel more comfortable, and are a reminder of normal home life. The cat's silky coat makes it more strokable than the rougher texture of a dog's coat, which is more likely to be patted.

At the End of the Day

While the sentiments on burial plaques to cats can sometimes bring a smile to the face, few people are so hard hearted that they cannot recognise the real feelings that reside there.

Putting a memorial over a cat's grave is not a new idea, for one of the most celebrated tombs was erected by Madame de Lesdiguières in the seventeenth century. On the mausoleum, the cat's likeness was carved in marble and placed on a marble pillow. This was at a time when the cat was beginning to be accepted at the French court. It tangibly marked that point in the history of the cat when in Europe it began to enter a new relationship with us and to change from being considered a necessary evil into a loved companion.

A gravestone to a cat in Los Angeles. Our relationship with our cat does not stop when contact turns to memory. In some cities in the USA it is illegal for pets to be buried in your own garden, and pet cemeteries are common. For housebound cats it can be the first and last time they leave their ownes' homes

Further Reading

Alderton, David. *Wild Cats of the World* (Blandford, 1993)

Armitage, P. and Clutton-Brook, Juliet. 'A Radiological and Historical Investigation into the Mummification of Cats from Ancient Egypt', J Arch Sc 8: 185–196 (1981)

Baldwin, James. 'Ships and the Early Diffusion of the Domestic Cat', Carnivore Genetics Newsletter, 4: 32–33 (1979)

Borchelt, Peter, and Voith, Victoria. 'Aggressive Behaviour in Dogs and Cats', Compendium on Continuing Ed 7: 11 (1985)

Churcher, Peter and Lawton, J. 'Predation by Domestic Cats in an English Village', J Zool Lond 212: 439–455 (1987)

Conway, William. *Dawn of Art in the Ancient World* (Percival, 1891)

Elton, C. 'Cats in Farm Rat Control', Brit J Animal Behav 1: 151–155, (1953)

Ewer, R.F. *The Carnivores* (Weidenfeld and Nicolson, 1973)

Karsh, Eileen. 'The Effects of Early and Late Handling on the Attachment of Cats and People', in 'The Pet Connection' Conference Proceedings, ed Anderson, R., Hart, B. and Hart, L. (Globe Press, 1983)

Leyhausen, Paul. *Cat Behaviour* (Garland STMP Press, 1979)

Loxton, Howard. *The Noble Cat* (Merehurst, 1990)

Gebhardt, Richard. *A Standard Guide to Cat Breeds* (Macmillan, 1979)

Macfarlane, A. 'Witchcraft Prosecutions in Essex 1560–1680', D Phil thesis (Oxford Univ, 1967)

Mead, Chris. 'Ringed Birds Killed by Cats', Mammal Rev 12:4 183–186 (1987)

Naville, Edouard. 'Bubastis 1887–1889', 8th Memoir of the Eygpt Exploration Fund (Kegan Paul Trench, 1891)

Seidensticker, John, and Lumpkin, Susan. *Great Cats* (Merehurst, 1991)

Simpson, Frances. *The Book of the Cat* (Cassell, 1903)

Stahl, Philippe. 'Le Chat Forestier d'Europe (*F.silvestris*): Exploitation des Resources et Organisation Spatiale' (Centre National d'Etudes sur la Rage et la Pathologie des Animaux Savages, 1986)

Robinson, Roy. *Genetics for Cat Breeders* (Pergamon Press, 1977)

Tabor, Roger. 'General Biology of Feral Cats' in *The Ecology and Control of Feral Cats* (Universities Federation for Animal Welfare, 1981)

Tabor, Roger. *The Wild Life of the Domestic Cat* (Arrow, 1983)

Tabor, Roger. 'The Changing Life of Feral Cats at Home and Abroad', Zool J Linnean Soc 95: 151–161 (1989)

Tabor, Roger. *Cats: The Rise of the Cat* (BBC Books, 1991)

Todd, Neil. 'Cats and Commerce', Scientific Amer, 237: 100–107 (1977)

Turner, Dennis, and Bateson, Patrick. *The Domestic Cat* (Cambridge Univ Press, 1988)

Weir, Harrison. *Our Cats and All About Them* (Clements, 1889)

Acknowledgements

The writing of this book has developed from around twenty years of my working as a cat biologist and behaviourist. Both for my researches and for television the cat has taken me to many countries around the world. This has enabled me to pull from a large well of information in writing the book, into which many people have kindly contributed. I cannot possibly thank here everyone who has been involved over the years, and I hope that anyone I cannot squeeze in will nonetheless accept my thanks.

Special thanks are due to Sue Hall, Susie Hallam, Sue Cleave, Piers Spence and everyone at David & Charles in producing this book as a happy collaboration. Also for the *Understanding Cats* series thanks are due to Jenny Cropper at the BBC, and Colin Tennant, John Bowe and my colleagues at Cats on Film/Avalon Productions. Thanks are also due to Dick Meadows and the BBC *Cats* team; Joan Moore and Malcolm Colchester at Cat World; the British Museum (Natural History) plus Ornithology section at Tring; the British Museum (Egyptology Dept); Cairo Museum; Dept of Antiquities,

Egypt; National Museum of Thailand; Cat Survival Trust; Zagazig University; Van University; North East London Polytechnic (now East London University); Universities Federation for Animal Welfare; Dudwa National Park, India; Pettrac; Colchester Zoo; TICA (New York and Los Angeles); the Cats' Protection League HQ staff and the CPL shelters at Battlesbridge and White Notley, especially Olive Hammond and Julie Kircher; Alley Cat Allies in the USA, particularly Louise Holton and Becky Robinson; Holly Hazard and the Doris Day Animal League; Wendy Horn and Manx National Heritage; Helena Sanders and all at Venice DINGO; Christine Pierson and Cats Assistance to Sterilize, Australia; Doris Westwood and the Fitzroy Square Frontagers and Garden Committee; Debbie Rijinders and Ineke Aldendorff of Stichting De Zwerfkat, Holland; Lord and Lady Fisher; Stuart Baldwin, Fossil Hall; Karen Heinick, Florida State University; Brian McNerney and friends; the London cast of *Cats* and the Really Useful Company; the veterinary surgeons Alan Hatch and Jenny Remfry; Alain Zivie; Roy Robinson;

Charles Llewelyn and family; Juliet Clutton-Brook; Major General Pengrpicha; Joan Hodge; Mike Jackson; Harry Bourne; Kutbettin Bahadir; Ann Baker; Peter Churcher; Rachel and the Cooke family; Hank and DD Tyler; Sarah King; Michael Harding; Robin Kiashek and Georgie Oldroyde; Ed and Malee Rose; Gay Frank; Bob, Vally and Charlotte Hudson; Jackie and Barry Wood; Solveig Pfluger; Dawn Gulliver; Mrs A. Wright; Wendy Thornton; Jean Murchison; Ros Elliot; Bernice Mead; Barbara Castle; Anne Bailey; Sharon McAllister; Mr and Mrs A. Gilbert.

Finally it gives me especial pleasure to record my particular thanks to Liz Artindale, without whose hard work and considerable support this book would have taken far longer to prepare, and whose photographic skills have much enhanced its appearance.

I have delighted in the acquaintance of many cats during my researches giving rise to this book – particularly my past cat Mr Jeremy Fisher, and my present pair Tabitha and Leroy.

ROGER TABOR 1995

Index

(Page numbers in *italics* indicate illustrations)

040-670-1